NOISE
IN AND AS
MUSIC

EDITED BY AARON CASSIDY AND AARON EINBOND

Published by University of Huddersfield Press

University of Huddersfield Press
The University of Huddersfield
Queensgate
Huddersfield HD1 3DH
Email enquiries university.press@hud.ac.uk

First published 2013
Text © The Authors 2013
Images © as attributed

Every effort has been made to locate copyright holders of materials included and to obtain permission for their publication.

The publisher is not responsible for the continued existence and accuracy of websites referenced in the text.

All rights reserved. No part of this book may be reproduced in any form or by any means without prior permission from the publisher.

A CIP catalogue record for this book is available from the British Library.
ISBN 978-1-86218-118-2

Designed and printed by
Jeremy Mills Publishing Limited
113 Lidget Street
Lindley
Huddersfield HD3 3JR
www.jeremymillspublishing.co.uk

Contents

Acknowledgements		vii
Contributors		ix
Introduction Aaron Cassidy and Aaron Einbond		xiii

Part 1: Theories, Speculations, & Reassessments

Interview	Ben Thigpen	3
Chapter 1	Black Square and Bottle Rack: noise and noises Peter Ablinger	**5**
Interview	Antoine Chessex	9
Chapter 2	Un-sounding Music: noise is not sound James Whitehead (JLIAT)	**11**
Interview	Alice Kemp (Germseed)	31
Chapter 3	Noise and the Voice: exploring the thresholds of vocal transgression Aaron Cassidy	**33**
Interview	Maja Solveig Kjelstrup Ratkje	55
Chapter 4	Subtractive Synthesis: noise and digital (un)creativity Aaron Einbond	**57**
Interview	Pierre Alexandre Tremblay	77

Chapter 5	Noise Music Information Retrieval Nick Collins	79
Interview	Eryck Abecassis	97
Chapter 6	Inside Fama's House: listening, intimacy, and the noises of the body Martin Iddon	99

Part 2: Practices

Interview	George Lewis	121
Chapter 7	"We Need You To Play Some Music" Phil Julian	**125**
Interview	Lasse Marhaug	129
Chapter 8	Beyond Pitch Organization: an interview with Michael Maierhof Sebastian Berweck	**131**
Interview	Kasper Toeplitz	145
Chapter 9	Materiality and Agency in Improvisation: Andrea Neumann's "Inside Piano" Matthias Haenisch	**147**
Interview	Franck Bedrossian	171
Chapter 10	Noise-Interstate(s): toward a subtextual formalization Joan Arnau Pàmies	**173**
Interview	Diemo Schwarz	189

Chapter 11	Molding the Pop Ghost: noise and immersion Marko Ciciliani	**191**
Interview	Ryan Jordan	209
Chapter 12	Qubit Noise Non-ference: a conversation Bryan Jacobs, Alec Hall, and Aaron Einbond	**211**

Appendix: Color Images   **229**

NOISE IN AND AS MUSIC

Acknowledgements

The editors would like to thank: Monty Adkins, Professor of Experimental Electronic Music, University of Huddersfield, for his support, encouragement, and guidance, particularly in the early stages of the project; Graham Stone, Information Resources Manager, Computing and Library Services, University of Huddersfield, who provided essential project support and served as the primary intermediary between the editors, the editorial board of Huddersfield University Press, and the book's printer, Jeremy Mills Publishing Limited; Hazel Goodes, Publishing Manager, and the rest of the staff at Jeremy Mills; Pierre Alexandre Tremblay, Professor of Composition & Improvisation and Studio Director, University of Huddersfield, for his assistance in contacting the interview contributors; Carter Williams, for his translation of the chapter by Matthias Haenisch; and, in particular, Tim Rutherford-Johnson, for his outstanding, thorough work as copyeditor.

Thanks are of course also due to each of the contributors to the book for their considerable investments of time, energy, and expertise, as well as the many friends and colleagues who provided both formal and informal review and editorial advice, and to those who additionally contributed papers, workshops, installations, and performances for the Noise In And As Music Symposium held at the Centre for Research in New Music (CeReNeM), University of Huddersfield, October 4–6, 2013.

NOISE IN AND AS MUSIC

Contributors

Eryck Abecassis is a Paris-based musician active as a composer of instrumental and electronic music, noise, and opera, and as a performer of electric bass/guitar, modular synthesizer, and laptop. His work can be heard in streets, landscapes, and urban architecture through concerts and installations worldwide.

Peter Ablinger was born in Schwanenstadt, Austria, and lives in Berlin.

Franck Bedrossian is a French composer and teaches composition at UC Berkeley. His works have been performed in Europe, in the USA, and abroad, most of them published by Editions Gérard Billaudot. CDs of his music have been released on Sismal records, Aeon, NEOS, and Ensemble Modern Medien.

Sebastian Berweck is a pianist known for his energetic interpretations of unusual repertoire in- and outside the piano and for his work with electronics. He is a member of stock11 and recently completed his PhD at the Centre for Research in New Music (CeReNeM), University of Huddersfield.

Aaron Cassidy is Reader in Composition and Research Coordinator for Music & Music Technology at the University of Huddersfield. His work can be heard on recordings from NEOS, HCR, NMC, and New Focus Records.

Antoine Chessex is a Swiss composer and sound artist whose works assume a wide diversity of forms spanning compositions for ensembles, solo performances, sound installations, and transdisciplinary projects.

Marko Ciciliani is Guest Professor for Electroacoustic Composition at the Institute of Electronic Music and Acoustics (IEM) of the University for Music and Performing Arts in Graz, Austria, and Lecturer at the University for Music and Performing Arts in Vienna. His music, which includes audiovisual compositions, electronic, and instrumental works, is regularly performed at festivals and concert series for contemporary and electronic music.

Nick Collins is Reader in Composition at Durham University. His interests include live computer music, musical artificial intelligence, computational musicology, and being a programmer-pianist.

Aaron Einbond is Research Fellow at the Center for Research in New Music (CeReNeM), University of Huddersfield. His work explores the intersection of composition, computer music, field recording, and sound installation, and can be heard on Carrier Records and Olive Music/Et'Cetera Records.

Matthias Haenisch is a musicologist and musician based in Berlin. He currently works as Research and Teaching Fellow at the Music Department of the University of Potsdam, Germany.

Alec Hall is a doctoral candidate in composition at Columbia University, studying with George Lewis. He is also a founder and director of Qubit, an experimental music organization in New York City.

Martin Iddon is Head of the School of Music and Professor of Music and Aesthetics at the University of Leeds. He is the author of *New Music at Darmstadt* and *John Cage and David Tudor* (both Cambridge University Press) amongst other journal articles and book chapters focusing on post-war new music.

Bryan Jacobs is a composer and guitarist currently residing in New York City. He is also the co-founder of Qubit, a new music initiative focusing on creative uses of technology.

Ryan Jordan is an electronic artist working with noise and self-built technologies to aid experiments in possession trance, derelict electronics, and hylozoistic neural computation. He runs nnnnn/noise=noise, a research laboratory and live performance platform in London for noise in art.

CONTRIBUTORS

Phil Julian is a UK based experimental sound artist/composer/musician. Audio works are available from labels including Entr'acte, Banned Production, Tapeworm, and Authorised Version.

Alice Kemp (Germseed) is an artist-musician based in South Devon, UK, with a background in self-taught experimental music, sound collage, and performance. She has been making collaborative and solo music/sound/performance work for two decades and has released works through Appliance Japan, Omcore Recordings, the Lazarus Corporation, and Fragment Factory.

George Lewis is Case Professor of American Music at Columbia University. A 2002 MacArthur Fellow, his compositions engage chamber, orchestral, computer, and improvised music, and his widely acclaimed book, *A Power Stronger Than Itself: The AACM and American Experimental Music* (University of Chicago Press, 2008), received the American Book Award.

Michael Maierhof is German composer of mostly chamber, vocal, and multimedia works that have been performed throughout Europe, the USA, Asia, and South America. Self-taught as a composer, he studied mathematics and music in Kassel and art history and philosophy in Hamburg.

Lasse Marhaug is a noise artist from Norway with over 20 years of activity behind him, including hundreds of releases and concert performances on all corners of the world.

Joan Arnau Pàmies is a doctoral student in composition and aural skills instructor at Northwestern University. His works are published by Periferia, BabelScores, and Éditions Bar&Co.

Maja Solveig Kjelstrup Ratkje is a composer and performer whose practice includes notated music, noise, free improvisation, electronics, and vocalization. She gives concerts and releases her music worldwide.

Diemo Schwarz is a researcher and developer in real-time music interaction at Ircam in Paris, composer of music based on noises from the environment, and improviser on his CataRT corpus-based concatenative synthesis software with gestural controllers.

Ben Thigpen composes music for loudspeakers (at GRM, EMS, VICC, etc.) and performs as an electronic musician. He lives in Paris and teaches computer music at Arts2 in Belgium. He has CDs on Sub Rosa and EMF Media.

Kasper Toeplitz is a composer and musician (electric bass, live-electronics) working between "academic" composition (orchestras, ensembles, opera) and electronic "new" or "noise" music—new forms of music for the 21st century and sheer electronic noise.

Pierre Alexandre Tremblay is a composer and performer on bass guitar and laptop, both as a soloist and within small ensembles. His music is released by Empreintes DIGITALes, Ora, and the Loop Collective. He is Professor of Composition & Improvisation and Director of the Music Studios at the University of Huddersfield.

James Whitehead (JLIAT) is a conceptual/drone/noise artist. He reviews noise for *Vital Weekly* and engages in independent work on noise theory, contemporary philosophy, and cybernetics.

Introduction

Aaron Cassidy and Aaron Einbond

> *Noise is a world where anything can happen, including and especially itself.*
> —Douglas Kahn[1]
> *Noise is music's dream of us.*
> —Morton Feldman[2]

The problem with noise is that it is everything. It is real, experiential, objective, measurable; it is also abstract, subjective, ambiguous, and contextual. It is both a thing and a relationship between things. It can be a state of communicative surplus, a chasm of ruptured communicative space, and also the material being communicated. It is potentially both overwhelming and reductive, both multiplicity and singularity. Every attempt at defining and discussing noise ends up somewhere in between these poles, bogged down between excess and incompleteness, avoidance and desire.

> At its most intense, noise music disorganizes the body; it disrupts the organization of the organs. It transforms the organs into a thousand ears, the ears into a vibrating, fluttering drum skin. Noise music addresses me as matter, rendering the body porous. I can feel it in my lungs, my stomach, my throat; it can turn me inside out.[3]

---

1 Douglas Kahn, *Noise Water Meat* (Cambridge: MIT Press, 1999), 22.

2 Morton Feldman, "Sound, Noise, Varèse, Boulez," in *Give My Regards to Eighth Street: Collected Writings of Morton Feldman*, ed. Bernard Harper Friedman (Cambridge: Exact Change, 2000), 2.

3 Marie Thompson, "Music for Cyborgs: the affect and ethics of noise music," in *Reverberations: the philosophy, aesthetics and politics of noise*, ed. Michael Goddard, Benjamin Halligan, and Paul Hegarty (London: Continuum, 2012), 211.

Or:

> But it is noise that we really understand. It is only noise which we secretly want, because the greatest truth usually lies behind the greatest resistance.[4]

Luigi Russolo's "The Art of Noises" is one hundred years old this year, and in many ways we are no further along in our understanding of noise or its role in artistic practice. Noise is still both other and ordinary, both manufactured and discovered, both provocation and invitation. It is still a mirror of the sounds of life, a confrontation and conversation with our industrial, ecological, cultural, and political surroundings. It is also still fetishized and misrepresented, an easy mask or smokescreen. And despite a century of musical-technological advances and ever more daring cultural practices that have augmented the noise lexicon, in the end "noise" is not ontologically different now than it was then. The questions that noise raises of its makers and its listeners are, at their core, the same as they were in the context of Russolo's manifesto.

What *is* different is the role noise plays as musical material. The language of noise as an artistic practice has changed in remarkable ways. Noise's vocabulary and syntax have been pushed further than Russolo could have ever imagined, its scale and magnitude far beyond what was conceivable in 1913.[5] But it is not simply specific "noises" that are now, in 2013, different—whether their sound, their character, their timbre, or even their intensity or density. In the most extreme (or perhaps more dangerously, the most "pure") examples of current artistic noise practice, noise's possible meanings have multiplied and expanded to the point of dissolution. The noise of Peter Ablinger, Merzbow, Dror Feiler, and others is not another node in an existing network of signification but instead a deliberate "de-differentiation," an obliteration

---

4 Feldman, "Sound, Noise, Varèse, Boulez," 2.

5 This includes a century of developments both artistic and technological, social and political, not least the fallout of Russolo's futurist colleagues' fascist sympathies (even if Russolo himself may have been opposed to them). See Luciano Chessa, *Luigi Russolo, Futurist. Noise, Visual Arts, and the Occult* (Berkeley: University of California Press, 2012), 8.

INTRODUCTION

of hierarchies and codes. This is often a music that foregrounds what JLIAT calls "data without information,"[6] a collapse into asignification. Freed from its earlier semiotic function as a symbol or cipher, noise now demands its own ontological rules, its own conditions for meaning in the context of the "big data" of our century. Our noise requires an entirely new way to process its sounds. That is, a new way to listen. A new way to *make*.

To that end, the central focus of this book is the practice of noise: noise in music, noise and music, noise as music. More importantly, it is about the state of that practice *now*. The principal texts in the field over the last several years have been primarily historical in their focus, whether Paul Hegarty's *Noise/Music: A History*, Douglas Kahn's *Noise Water Meat*, or Jacques Attali's *Noise: The Political Economy of Music* (now a classic). Instead, our aim was to engage current practitioners, to expose a cross-section of the current motivations, activities, thoughts, and reflections of composers, performers, and artists who work with noise in all of its many forms. Even the chapters that are more musicological in nature are written by musicians engaged with noise practice as composers or performers. The book is about noise as material: noise as a raw building block of music, noise as method of composition, noise as a way of conceiving musical notation, ways of evaluating or codifying noise, ways of using noise to better understand the act of listening.

Origins

The story of the development of this book stems from two sources. One is the Noise Non-ference organized by Qubit in New York in March 2013. As retold in the final chapter of the book, what began as a modest call for participation in an event somewhere in between a short conference and a miniature music festival received overwhelming interest from applicants

---

6 James Whitehead, "Two noises," accessed August 5, 2013, http://www.jliat.com/txts/Two%20Noises.pdf.

around the world. It seemed noise was a topic with particular resonance, with a much wider reach than had been anticipated. We felt that noise in its many possible meanings, by many different practitioners, required further exploration.

The other seed of the idea came from the postgraduate seminar series in contemporary music at the University of Huddersfield, which each year focuses on four or five thematic strands. Noise was one of the topics we proposed for the 2012–13 academic year, but this quickly snowballed to include a conference, the conference quickly became a much more diverse and multidisciplinary symposium including concerts and installations alongside papers and presentations, and an initial proposal of a publication of the conference proceedings quickly turned into this book. Critically, though, the ethos of the book comes from the motivations behind the postgraduate seminars—the topic was proposed not because we felt we had things to *teach* about the notion of noise, but instead because we had questions to which we did not know the answers, propositions and curiosities we wanted to explore, repertoire that we wanted to study but that we did not feel we had the vocabulary or analytical methodology to describe or explain adequately. The process was an investigative, theoretical one. We wanted to delve more deeply into the notion of noise as material—noise as a way of thinking and expressing musically—and to invite as many voices as possible to join in the investigation.

Structure

The book is organized into two halves, though in truth several of the chapters could have appeared in either. The first approaches noise largely from a theoretical perspective, proposing ways of thinking about, perceiving, evaluating, and comparing noise. These chapters examine the origins of musical noise, the thresholds of noise, and more generally the questions that noise raises, particularly as musical material. Most broadly, this half of the book is about listening to noise, whereas, on the whole, the second half is about making it.

Part one consists of theoretical proposals by six composers that, while principally about noise in general or about the noise practices of others, reflect back on the authors' own creative work in intriguing ways. Peter Ablinger, who is perhaps more engaged with the question of noise than any composer of his generation, describes two streams of noise, one a framing and recontextualization analogous to the "ready-mades" of Marcel Duchamp, and the other the totality of noise—akin to the monochromatic paintings of Kazimir Malevich or Ad Reinhardt—and the projective, active listening it engenders. James Whitehead (aka JLIAT)—an artist whose conceptual music pushes at (and often usurps) the boundaries of noise and music—extends the notion of totality even further, including an examination of the ways in which the boundless space of digital music requires a reassessment of the relationship between music and sound. Our own chapters, respectively, include a proposed definition of the thresholds of vocal noise that draws on notions of transgression, violation, or breech, and an examination of subtractive synthesis, concatenative synthesis, and Music Information Retrieval as artistic methodologies parallel to the poet Kenneth Goldsmith's notion of "uncreative" creativity. Nick Collins extends Music Information Retrieval methods further as an analytical tool to quantify and compare noise content in the work of Merzbow, Masonna, and others. And finally, Martin Iddon turns the focus inward, exploring the noises of rumor and the metaphor of Fama's house to locate the site of listening in the intimate privacy of a corporeal, visceral "listening body."

The second part of the book focuses on noise practice from a more personal, self-reflexive, and often autobiographical standpoint and aims to provide snapshots of the role of noise in various geographical and stylistic communities. The contribution of British noise performer Phil Julian includes an account of the role of individuality and the standardization of noise music as a genre and a form. The pianist Sebastian Berweck and the composer

Michael Maierhof, both members of the stock11 collective based in Germany, discuss instrumentation, performance practice, and compositional technique in non-pitched music, including a discussion of what constitutes an "amateur" performer in the case of the unfamiliar sound-production techniques regularly found in Maierhof's work. Also from Germany, musicologist and performer Matthias Haenisch examines the instrumental practice of the Berlin-based improviser Andrea Neumann interpreted through Bruno Latour's Actor-Network Theory. The Catalan-born, USA-based composer Joan Arnau Pàmies explains the role of noise and information theory in his notational practice in works for acoustic instruments, including discussions of the distortion and upending of stable relationships between duration, physical movement, and notations of time and space. And the Austrian composer Marko Ciciliani writes about his work with the dense superimposition of pop songs in his electronic sound installation *Pop Wall Alphabet*, and provides a textual corollary that mirrors the work's compositional methodology through a noisy poetry. Part two concludes as our project began, with a discussion between the three organizers of the Qubit Noise Non-ference examining the process of curation as it reflects on the current noise music scene in New York and beyond.

Interviews

One of the book's unique features is a collection of answers to a simple, two-question "interview":

What is noise (music) to you?

Why do you make it?

We solicited responses from a diverse array of practitioners, some of whom identify themselves as "noise musicians" and many who do not. The twelve musicians who contributed answers to the interview include composers of both notated and electronic music, performers and improvisers on acoustic instruments and electronic ones, sound artists, vocalists, curators, software programmers, makers of dense, loud sounds and intimate, sparse ones. The variation in how contributors interpreted and responded to these questions complements the variety in their sonic approaches. Their responses are at times candid, elegant, funny, revealing, impassioned, and even dismissive (though we admit to being slightly disappointed that none of the musicians we approached responded with a simple "fuck off").

The interviews are interspersed between the book's chapters and serve as interludes, pivots, and portals. In some cases they have a grounding function, returning us from the theoretical and the speculative back to the experiential and the autobiographical. At other times they are poetic, personal epigrams that instead spiral outward, away from measured efforts to define and delimit and explain.

Definitions

In all, the contributors' compelling answers suggest self-definition as a key characteristic common to recent noise work. As expected, many contributors distanced themselves from genres, scenes, and common practices, just as they may distance their own musical identities from clear-cut labels like composer, performer, improviser, or programmer. The fluidity of noise-making is revealed as one of noise's most appealing features, and it is perhaps this malleable, shape-shifting erasure of boundaries and roles that is most relevant to our current climate. George Lewis refers to "the connection between noise, improvisation, spontaneity, and nomadism," summarizing an image of migration common to several texts in the book. By bringing these forces into conversation through the following pages, we hope to accentuate a flow of ideas and sounds that few other musical topics could provide.

Perhaps surprisingly, there was little disagreement or debate from the various contributors across the book about what the term "noise" might mean. There seems to have been a certain consolidation and consistency of definitions of noise both as a concept and as a practice. However this does not mean there is a consensus. It seems that noise is a sort of Rorschach test for many practitioners' recent work, which speaks to many yet which everyone hears in a different way. While some of the contributors nod to standard definitions (whether from the domains of acoustics, signal processing, information theory, or politics), the diversity of possible hearings is what gives noise its current power and relevance.

What is interesting as well is that noise, for these contributors, is not an anti-art, the self-defeating genre that Cécile Malaspina is probably referring to when she says, "As a proper name, 'noise' comes down to the absurd designation of nothing much: the blandness of a generic negation."[7] Or, as Douglas Kahn puts it, "noise is always subject to operations that render it nonexistent."[8] The act of making noise as art risks undermining its noisiness, and yet the art of noise practice continues to grow. This dilemma is acknowledged by Pierre Alexandre Tremblay's term "post-noise." Like post-glitch, it is a practice now removed from its de-familiarized origins, with its own developing vocabularies, syntaxes, social structures, and performance practice. It is this trend that excited us most as we compiled the contributions to the present book. With noise no longer defined primarily through its "otherness," it is time to put it center-stage and let noise and its makers speak for themselves.

---

7   Cécile Malaspina, "The noise paradigm," in *Reverberations*, ed. Goddard, Halligan, and Hegarty, 60.

8   Kahn, *Noise Water Meat*, 26.

Part 1: Theories, Speculations, & Reassessments

Benjamin Thigpen

What is noise (music) to you?

Noise is a manifestation of truth.

Why do you make it?

To be in touch—in an immediate and total connection—with the real.
To participate in the opening up of what is undisclosed and unexplained.
To be drawn into a vortex of primal energy …

NOISE IN AND AS MUSIC

Black Square and Bottle Rack: noise and noises

Peter Ablinger[1]

*Some day every artist
has to choose between
Malevich and Duchamp.*
—Ad Reinhardt[2]

Noise and noises are not the same. In fact, they can be almost opposites. What the singular form refers to is the totality of white noise. What the plural refers to is the many individual objects, the event-related noises of everyday life. Obviously this distinction is very important to my own work and the use of "*Rauschen*" (white noise) within it, but what I want to argue here is that this distinction has already achieved a historical dimension—although it is one that seems not to be widely recognized.

As is so often the case, the visual arts discourse on the matter is decades ahead of the music discussion. The Ad Reinhardt quotation above is now 45 years old. To arrive *in medias res*, I want to equate Duchamp's readymades (e.g., the bottle rack) directly with the individual noises of everyday

---

1 Additional editing by Henry Anderson

2 The drawing is from my notebooks, circa 1997. Above the drawing I wrote (back-translating here from German): "*Ad Reinhardt 1967: 'Basically, in the 20th century there is only the choice between Malewich and Duchamp.'*" Another quotation of Reinhardt, as found online, says: "One must decide between Duchamp and Mondrian." [Editor's note: the version above is quoted in Phong Bui, "Mel Bochner with Phong Bui," *The Brooklyn Rail*, May 9, 2006, accessed July 19, 2013, http://www.brooklynrail.org/2006/05/art/in-conversation-mel-bochner-with-phong-bui.]

life, and to connect Malevich's most abstract paintings (like the black or the white square) to the idea of totality and the sum of all sounds—which, by definition, is white noise.

As I read Russolo and perceive Cage, their noises or *rumori* are about sounds as individual acoustic events, as the material, the building blocks, the modules (or found objects) that might constitute a composition. Thus, noises also represent the equivalent and complement of tones or instrumental sounds. In each of these respects, white noise is the opposite. White noise/ *Rauschen*[3] is not an individual in the sonic world, but its suspension. It is not an equivalent or a complement of tones, but rather it *contains* both the tones and the noises. It is the totality of all sounds and noises, their sum.

When John Cage, in "Lecture on Nothing," talked about Debussy, and about removal as a principle of composition, he appeared—at least for that moment—to be close to the idea of *Rauschen*: "Somebody asked Debussy how he wrote music. He said: I take all the tones there are, leave out the ones I don't want, and use all the others."[4]

I don't know if Cage ever returned to that. After all, he had something against (acoustic) totalities, against Xenakis, against Free Jazz, and against situations in which the individuality of discrete sounds would be suspended in a mass or sum. And here we are now, exactly at this point of distinction— and in addition, in our hands we hold the key for opening up and acquiring its historical dimension.

The bottle rack motif, in my argument, does not refer to the ready-made strategy itself (in the sense that the state of readiness is not further overworked during the process of becoming a piece of art) but more to the techniques of acknowledging noises as sonic individuals. Usually this means isolating, framing, recording, de-locating sounds (or objects), taking them

---

3 Significantly, the German word *Rauschen* has no plural. That is probably why I—also in English—tend to differentiate between the singular and the plural form: noise versus noises.

4 John Cage, "Lecture on Nothing," in *Silence* (Middletown: Wesleyan University Press, 1961), 117.

out of their "natural" surroundings—in which they cannot be recognized as "individuals"—and repositioning them in an art or music context.

Kurt Schwitters, in 1920, wrote precisely about this aspect of removing something from its context: "the artist recognizes that in the world of manifestations that surround him, any particular only needs to be bordered and torn out of its context to result in a work of art."[5]

"Liberation of sounds" (Varèse, Cage) is necessarily connected to the techniques of isolation and de-contextualization. On a more philological or abstract level, however, I wonder whether the rhetoric of liberation is hiding something: the intended individualization—only subjects can be liberated!—is in truth an objectification.

Russolo, Varèse, Cage, *musique concrète* (Pierre Schaeffer), *musique concrète instrumentale* (Helmut Lachenmann), soundscapes, and field recordings (R. Murray Schafer)—all these and many more are thus clearly related to the bottle-rack aspect, or to (individual/isolated) *noises*.

It is the black square aspect of music, however, that is the less explored, the less exposed, and that must be treated carefully. I believe, though, that history itself has already delivered enough reference points to indicate the black square's relevance and true existence.

Cage's notion about Debussy has been our starting point for this, and Debussy himself could be observed more closely in that light. Josef Matthias Hauer is another early modernist dealing with the idea of totality. Yves Klein's *Symphonie Monoton Silence* relates to his monochrome paintings and therefore also to Malevich. Xenakis, whose interest in mass phenomena stands in opposition to Cage's individualization or liberation of sounds, provides another example. Certain qualities of extreme density in improvised music (Cecil Taylor) or in Noise Rock could also be discussed, and I am quite sure there are further examples, especially from the last 30 years.

---

5 In German: "Der Künstler erkennt, dass in der ihn umgebenden Welt von Erscheinungsformen irgend eine Einzelheit nur begrenzt und aus ihrem Zusammenhang gerissen zu werden braucht, damit ein Kunstwerk entsteht."

Additionally, I present a short list of earlier examples of the use of white noise in different art pieces. Significantly, more artists in this list are media artists than musicians:

Peter Kubelka, *Arnulf Rainer*, 1960
Nam June Paik, *Zen for Video*, 1962
William Anastasi, *Microphone*, 1963
Howard Jones, *Air 44*, 1970
Marina Abramović, *Sound Ambient White Noise*, 1973
Wolf Kahlen, *Drop Outs*, 1993

I hope my notes so far have provided enough keywords to initiate discussion about the two streams of noise(s), and also to give some sense of the diversity (if not antagonism) of the two general concepts, each of which lead to different and often opposite artistic and compositional strategies.

But I would like to add one further thought related to my own research and to the "totality" aspect of white noise, because it is exactly this aspect that can generate effects of high "individualization." As soon as we shift our attention to its perceptual consequences, as soon as it is no longer about treating the sounds as individuals to be liberated, but about the *real* individuals—about us, the listeners—then white noise becomes a wonderful field for experience and exploration. In particular, the field of (individual) projection, interpretation, and acoustic illusion is well suited for examining the area of listening and the constructive role of our brain in that process.

What I learned from my own work—and especially its black square aspect—is that listening has nothing to do with an outer world that we receive passively. Rather, listening is a creative activity that forms both what we hear and how we hear. We are creating, therefore, nothing less than ourselves.

Antoine Chessex

What is noise (music) to you?

Noise is Sound, Sound is Music, Noise is Music. The vibration of the air.

Why do you make it?

Because I'm a composer and a performer, so working with sound (or noise or music) is what I do.

I'm not preoccupied at all by genres or categories to classify music (or noise or sound) and can't conceive a hierarchy between (for instance) a dissonant cluster played by a string quartet, loud feedback created by a microphone plugged into a guitar amplifier and a distortion pedal, or a field recording of a densely populated city at rush hour. Those are simply available sounds (or noise or music) with different characteristics.

NOISE IN AND AS MUSIC

Un-sounding music; noise is not sound

James Whitehead (JLIAT)

SPIEGEL: And what takes the place of philosophy now?
HEIDEGGER: Cybernetics.[1]

The provisional title for this chapter was "Noise is more than sound," which is unsatisfactory for a number of reasons. Firstly, it can bring to mind the idea of set theory and Venn diagrams with circles labeled "Noise," "Music," and "Sound" in various configurations, allowing various understandings of "noise," "music," or "sound," though always as bounded, finite objects in fixed and finite relationships to one another. Secondly, the framework in which this appears is "Noise in and as Music," which again makes noise into some object, and an object captured "in" music, as if noise is something like a wild animal, a tiger or lion to be caged in a circus. Further, the word "in" is also a culprit, as the big game hunter or specimen collector seeks to put noise into music and noise is effectively tamed or killed in the process. Hence the change in title.

So if noise is not sound, what is it? Two meanings of "sound" are both simultaneously in play here, however now this is not a circle on some diagram, neither is it a beast, nor an object, for I propose that it is not an "it" at all and so even the great metaphysical hunter Heidegger will not find anything, any thing, to cage. (Pun intended.) Noise, noise *qua* noise, is something like infinity; it is neither fixed nor totalizable. Music's relation to it can be understood as being similar to mathematics and infinity, where infinity is the territory or space that mathematics inhabits. Noise is the territory inhabited

---

1 Interview given in 1966, quoted in Frank J. Tipler, *The Physics of Immortality* (London: Macmillan, 1994), 86.

not only by music but by all representation, all signification. The inhabitants are finite signifiers, capable of an infinite play of meaning, though they can make no difference to noise's infinite space: subtracting infinity from infinity leaves us with infinity.

Meaning, no Meaning

We have emerged from a period in the history of theory defined by the linguistic turn, in which definitions play in language games and authorities of meaning, authors, and Gods have been reported dead. Signifiers are arbitrary; grammatical structures collapse and re-form as opposites in texts; musical and artistic structures no longer have meaning but use, a use that has become increasingly entertainment, an entertainment for us. The failure of the philosophy and art of the recent past was the failure to locate any absolute. A continual self-examination, self-doubt, and self-questioning, in which music became a series of experiments, resulted in the abandonment of meaning for the pragmatics of use and a period typified by paranoia of meaninglessness that collapsed into the assured, sensational, empty presence of postmodernity's schizophrenia.

No such disease can be found in science during that period and even through postmodernity, which might account for contemporary philosophy's interest in science as the source of mathematizeable truths. Science did not look inward at its structures but created new ones in its exploration of the great outdoors.[2] Despite the obvious structures in the great achievement of Western music, these structures were never sufficient, never necessary, and so became a victim of metaphysical critique.[3] Therefore, it might be argued that,

---

[2] I am referring to relativity theory replacing or adding to Newtonian mechanics, quantum theory, and more recent work on field theory of M-Theory and Strings.

[3] For example, "Non-conforming music has no defense against the indifferentism of the mind, that of means without purpose." Theodor Adorno, *Philosophy of Modern Music* (London: Continuum, 2007), 15.

from Kant onwards, philosophy led art down some garden path to oblivion.[4] The problem of meaning in the arts became very personal, to the extent that its present state could be criticized as nothing other than a cult of personality. Being personal for a moment, I did not engage in that. Because of a quite unexpected set of events, I turned my attention to computer science, where definitions are fixed.[5]

"Noise," a Dictionary Definition

1. a. Sound or a sound that is loud, unpleasant, unexpected, or undesired.
   b. Sound or a sound of any kind: The only noise was the wind in the pines.
2. A loud outcry or commotion: the noise of the mob; a lot of noise over the new law.
3. Physics: A disturbance, especially a random and persistent. disturbance, that obscures or reduces the clarity of a signal.
4. Computer science: Irrelevant or meaningless data.[6]

It should be clear that definition 1a allows for an infinite variety of personal expression and opinion, and much has been gained by its exploitation. Postmodern art has been typified by a general destratification, a heterogeneity

---

[4] The path to oblivion or nihilism, the philosophies of Kierkegaard, Nietzsche, Heidegger, Sartre, Camus, et al. It is a nihilism still to be found in contemporary philosophy, for instance in Ray Brassier, *Nihil Unbound* (Basingstoke: Palgrave Macmillan, 2007). Compare this to science in its predictive achievements: for example, the periodic table as it developed showed clearly where undiscovered elements should fit, and as such their structure and properties could be known before any empirical discovery, one of the high positivist moments in science.

[5] No one event in thinking about music and theory made me examine the binary structure of sound files, but a question regarding all pitches combined on an internet board first led me to examine the structure of WAV files in storing sound. A career in computer programming and systems analysis provided the necessary knowledge and access to tools, and then reading the remarkable assertion by Heidegger quoted at the opening of this chapter led me to investigate the world of binary data from a musicological/philosophical perspective.

[6] http://www.thefreedictionary.com/noise, accessed February 20, 2013.

of "what ever it means to you is what it means." Noise artists are often characterized by or deliberately make themselves predicated on such a definition of noise, and serve to critique and challenge conformity in the arts and society, even to the extent of ironically challenging themselves. Such expressive freedom, for good or bad, rapidly regresses into itself as a kind of noise. Definition 4 is useful as it allows us to absolutize music by showing that a definition of noise is not possible, because no satisfactory containment is possible for a procedure that is necessarily not containable in any finite structure. Such an absolute and deterministic process arises from the nature of noise in computer systems, which are not subject to the inability of fixing a meaning, as occurred in modernity, or to postmodernism's relativism of sensation.

Much of the computer science below is widely known, so I apologize for its pedantic exposition. We do not require a detailed description of the subject except to give us necessary terms for "identity" and "noise" without recourse to esoteric metaphysics or matters of opinion and taste. We need these definitions in order to attempt an objective reexamination of the terms, and from this a radicalizing of the ontology of music.

A simple computer (processor) can only add. It cannot do multiplication, as rather than guess what 7 x 8 is, it simply adds up 8 lots of 7. Furthermore, a simple processor cannot subtract. It achieves subtraction by using complimentary arithmetic, and this process is how a computer system can "identify" or recognize. Complimentary arithmetic only sounds complex. Here is how you compliment a binary number, for example 0100101:

Step 1. If you see a 0 change it to a 1 and if you see a 1 change it to a 0.
Step 2. Add 1.

That is it.

So
    00100101
Becomes
    11011010
Add 1
    11011011

Now we can perform subtraction by addition. For example if you want to subtract 37 from 52, depending on how you were taught, you will be carrying and borrowing; you will know how to subtract 7 from 2 and 4 from 5. You know many rules the computer does not. You can manipulate the digits 0 through 9. Our processor can only add 0 to 0, 1 to 0, 0 to 1, and 1 to 1. That is all it needs.

Here is the "subtraction" of 37 from 52 using complementary arithmetic:

00110100
11011011 +
―――――――
00001111

00001111 is 15 in decimal. If you do not know why or how this worked (and it did; 52 – 37 is 15), do not worry; neither does any computer.[7] When a computer system does something like recognize a face, it (more or less)

---

[7] Why the computer only performs addition and nothing else is part of the "keep it simple stupid" methodology of computing. If you can add, subtract, multiply, and divide simply by addition, then only one circuit for addition is required. Real computer systems may use other additional methods, but they still use Boolean logic that is very simple. Multiple processors and all the other technologies just speed up the process; they do not make it ontologically more sophisticated.

Here is how this "subtraction" is performed: in a 2's Complement number the highest bit is negative.

-128	64	32	16	8	4	2	U	
0	0	1	1	0	1	0	0	= 32 + 16 + 4 + 2 = 52
1	1	0	1	1	0	1	1	= -128 + 64 + 16 + 8 + 2 + 1 = -37
+								
0	0	0	0	1	1	1	1	= 8 + 4 + 2 + 1 = 15

In this addition the high bit to the left is "lost." On compact disks and in typical sound files the data is held as signed 16-bit integers. This gives a range of +32767 to -32768. If a sound file contains a set of numbers all the same, no change in value occurs, which is in effect silence. There are therefore 65,536 possible silences in this format.

turns the data into binary and subtracts. The logic is simple. If the result of subtracting two numbers is zero, they were the same; any other result tells you they were not the same. This is the cybernetics of identity or recognition!

Often, when you log in to a computer system, a process needs to check if your user ID is the same as the one stored to allow you access. The system needs to recognize your login identity. Imagine your ID is not something like James123, but "J." The computer subtracts the binary value for "J" from all its stored user IDs until it gets a result of zero. J is stored as 1001010 (ASCII 74), so the subtraction takes place as above. If it subtracts your ID from every ID it has and never gets a zero, it will tell you that you do not exist in its system. For real IDs of more than one character it subtracts each in turn, again "recognizing" your identity when it gets a zero result. Although we seem far from definitions of "noise" and "music," we have now a very simple definition of identity. A thing is identical to itself if its negation added to it leaves nothing. Suddenly Heidegger is back in the room, perhaps.

> SPIEGEL: And what takes the place of philosophy now?
> HEIDEGGER: Cybernetics.

Von Neumann Architecture (or "illegal" operations)

The Von Neumann Architecture is how modern computers work. One of the features of this architecture is shared memory. The computer's memory (RAM) is used by data, addresses, and instructions. Data is a picture you are drawing, a text file you are editing, a music file, etc. Addresses are pointers the computer uses to find "stuff" in memory and take data there. (Think of an envelope with data inside and an address on the outside.) Instructions tell the computer what to do; they are low-level machine code. For example, "take these bits here, fetch some bits from there." The bits could be data, the here and there places in memory. The point is, rather than have three types of memory—one for data, one for addresses, and one for instructions—they all share the same space, for technically good reasons. (Keep it simple, stupid.) Though this simple idea is efficient, problems can occur when data, addresses, and/or instructions get muddled.

A "typical" section of code might look like this:

| Instruction | Data | Address |
| Do this | to this | from here or send it here |

Each would consist of Bytes made of Bits of fixed length (typically 8, 16, or 32).

0101001010010000   11100001101010110   100101010001010

However, there is no way of knowing what 11100001101010110 means in isolation. If it is in the first part, it is an instruction; in the middle it is data; and if it is at the end, it is an address. The computer treats whatever arrives in its processing unit as such. If a bit goes missing then

0101001010010000   11100001101010110   100101010001010

becomes

101001010010000 1   1100001101010110 1   00101010001010 +?

Now the instruction starts with a 1 and uses the first bit of its data section to complete the instruction. The data now has a missing bit at the beginning and uses the first bit of the address. If the "new" instruction does not exist, 1010010100100001 is not recognized and the computer will not be able to continue. This is an illegal instruction! If the "new" address is not correct, it is an "illegal" address. If the data is altered, then the text file or MP3 or picture will have nonsense or rubbish at that place. Here is a simpler example using text.

TAKE THIS HERE

>TAKE = Instruction or operation code
>THIS = Data
>HERE = An address in memory

We lose the first part of the instruction, so now have:

AKET HISH ERE?

The T is lost, but the T from THIS is used to make up the gap, and so on. AKET is not a task we can perform, HISH is not recognizable data, and ERE? is not a recognizable place. We have nonsense, chaos, noise. This is how errors of "executing data" that are "executing code from a non-executable memory region" can occur.

By way of proof, an actual example. Below, the Microsoft Explorer program crashed trying to write data to a place in memory. This caused an "exception," a problem that the program could not handle, and so the operating system—in this case Windows NT—gave the following message:

Unhandled exception at 0x0a153739 in explorer.exe: 0xC0000005: Access violation writing location 0x090afc40

If we look at the place in explorer.exe where this occurred, we see:

Address	OpCode Meaning	English
04893739	movd dword ptr [edi+edx],mm2	move some stuff
0489373D	psrlq mm2,20h	shift some bits left
04893741	movd dword ptr [edi+edx+3],mm2	move some stuff

The example above shows memory in which an opcode, an instruction, tried to move some data into an area of memory it should not. Either this address in RAM was not there,[8] or this process (program or application) did not have the rights to perform this move (the location was being used by another process, for instance). Once the context of the bit string is lost, it effectively becomes noise. Such noise causes the program to fail. Coherence, structure, and meaning are lost.[9]

Such a definition of noise from within computer science defines noise as essentially destructive and not as some effect or affect.

The Correlationism of Music

One of the features of a certain group of contemporary philosophers—labeled variously as Speculative Materialism, Speculative Realism, Object Oriented Philosophy (OOP), or Object Oriented Ontology (OOO)—has been a critique not only of postmodern and continental philosophy but also of Anglo-American analytic philosophies, and the identification of

---

8 By "not there" I mean that the address generated was larger than the address space of the computer. If you have a street of 100 houses, you cannot send a letter to number 145.

9 This might clarify some problems with interpretation in the arts. Saussure established that in language signs are arbitrary, and this was further developed by Derrida in *Of Grammatology* to argue that language worked as a play of differences without any one or set being privileged. There is nothing in the letters "C" "A" "T" that is catty: CAT and DOG could be interchangeable for what they signified. Or in Albanian QEN and MACE: which is CAT and which is DOG? In effect, it has been said we are all speaking "dead languages" (Brassier, Lacan, etc.). There is nothing inside the text to give meaning (and the text here could be an image or a sound structure). Further, it has been argued there is nothing outside of the text that can provide it with a fixed definitive meaning. Famously, "il n'y pas de hors-texte," (Jacques Derrida, *Of Grammatology* (Baltimore, MD: John Hopkins University Press, 1976)), by which there is no arbiter of a fixed and final meaning of a text—neither the author, nor a psychoanalytical or metaphysical theory of interpretation, nor God. "The semantic horizon which habitually governs the notion of communication is exceeded or punctured by the intervention of writing, that is of a dissemination which cannot be reduced to a polysemia. Writing is read, and 'in the last analysis' does not give rise to a hermeneutic deciphering, to the decoding of a meaning or truth" (Jacques Derrida, *Limited Inc* (Evanston: Northwestern University Press, 1988), 20).

"correlationism" within all of these philosophies.[10] Their activity is closely associated with the arts, as was the continental school that they critique. The idea of correlationism effectively critiques all philosophy since Kant, and identifies Kant not as the originator of a Copernican revolution in thought but rather as a reactionary Ptolemaic in his unwillingness and prohibition against thinking "The Real."[11] The justification for this critique is a debate to be had in philosophy and not here. However, I want to employ the same correlationist critique of philosophy to music. Philosophical correlationism can be summarized as the idea that philosophical thought, properly metaphysical thought, never has access to the real, to things in themselves, but it has access to—in fact it is the correlation between—thought and its object. It is only in the correlation that we can ground a philosophical necessity, an absolute and objective knowledge.

> Thoughts without content are empty, intuitions without concepts are blind. The understanding can intuit nothing, the senses can think nothing. Only through their unison can knowledge arise.[12]

It is a gross simplification to say we only experience our perception and never what exists outside our perception and therefore cannot know anything of objects but only have knowledge of our experience of them. Key to human experience are Time and Space, but these are, in correlationism, not real

---

10 This field originated as "Speculative Materialism" from a conference held at Goldsmiths College, University of London, in April 2007. As the numerous titles indicate, the members of that conference (and others) are not so much a "group" or "movement" but instead philosophers who have an interest in a metaphysical realism as critique of the dominant forms of post-Kantian "correlationist" philosophy. For the sake of convenience I use the term "Speculative Materialism" throughout this chapter. The original conference members were Ray Brassier, Iain Hamilton Grant, Graham Harman, and Quentin Meillassoux.

11 "[T]he central notion of modern philosophy since Kant seems to be that of correlation. By 'correlation' we mean the idea according to which we only ever have access to the correlation between thinking and being, and never to either term considered apart from the other" (Quentin Meillassoux, *After Finitude* (London: Continuum, 2008), 5).

12 Immanuel Kant, *Critique of Pure Reason* (London: Penguin, 2007), 86.

things but merely the necessary constructs for us as humans to experience both the outside world and our inner consciousness.

> [T]ime is nothing but the subjective condition under which alone all intuitions can take place within us.[13]

Speculative Materialism wants to reject this correlation in favor of access to the Real, which science seems to have enjoyed, unlike philosophy, since Kant. One of its main motivations is combating the relativism of postmodernity, a relativism similar to the first dictionary definition above of noise as a matter of taste and opinion.[14] Regardless of philosophical correlationism, it appears so obvious that art exists as a form of correlationism that the idea is often ignored, for correlationism appears to be an *a priori* necessity that constitutes what art is. One of the consequences of a radical regard to noise and its relation to music is that it exposes and breaks that thought. Alternatively, noise can safely be regarded as one more trope or ingredient for music, perhaps even a dangerous supplement, but one that does not question music's ontology. Crucially, though, if this ontology is questioned then not only is music radicalized but also all the arts. It is from the radicalization of music by noise that a general radicalization of representation might be achieved.

There are numerous theories of art; however, most if not all posit an object and a subject, with the status of the object's art-ness located not in the object but instead in the nature of the relationship between the object and subject.[15] Music is heard, paintings are seen. Music is played, paintings

---

13 Kant, *Critique of Pure Reason*, 69.

14 "[M]athematics' ability to discourse about the great outdoors; to discourse about a past where both humanity and life are absent" (Meillassoux, *After finitude*, 26).

15 Typically, the subject is human, but this is not necessary. Birds, whales, and other sentient creatures may and probably do engage in music, according to some theories, but the engagement is still essentially the correlation between a subject and an object. Furthermore, musicians have historically found such "musics" of interest, including the more recent genre of music using field recordings.

are painted. Duchamp's urinal or Cage's silence exist in and because of this correlation. In fact they expose it, rely on it, and work with it. In Duchamp, the context provides the status of art. Cage's *4'33"* presents the impossibility of silence as an impossibility for us in our experiential relationship within the performance. Philosophically, the tree falling in the forest may or may not make a sound. No such dilemma exists in *4'33"*—there will always be sound. It is no surprise when Meillassoux posits a time before human thought and a time after. When likewise he proposes existence prior and post *Homo sapiens*, even the idea of existence and temporality where no cognition exists at all seems acceptable. However, if we move the claim from philosophical ontology to music, we arrive at something quite radical and contrary to the Cageian, correlational ontology of music. Music where no human exists to create the correlation is a very radical idea of music; the idea of non-correlationist music might appear impossible or absurd. We may think of the possibility of noise outside of a human correlation and outside of cognition: it may be debatable but it would not be ontologically impossible, otherwise no debate could take place. Therefore the acceptance of noise as music can either simply ontologize and make noise part of a human correlation or radicalize music itself by destroying its ontology. A choice needs to be made, but before it is, it is crucial to explore just what is at stake.

It would, I think, be difficult to maintain absolutely that sound is only a correlation. To propose that music existed before human thought and will exist after is much more contentious because music is accepted as the correlation in which humans must take a part.[16] Cage's silence is an impossibility for us, but not an impossibility *per se*. We have a precise analogy between Kant's

---

16 Sounds are vibrations and exist in much wider spectra than human hearing. Music may be distinguished only by having air as a medium. Though sounds are found in other media such as water or parts of the electromagnetic spectrum, they are vibrations that are impossible for humans to hear and are independent of any medium. To define sound as only that which humans can hear seems very anthropocentric at minimum and fraught with problems too extensive to explore here. Even if sound is so restricted as to require a medium, "noise" cannot be. Noise occurs in radio transmission and communication and computation. There is cosmic noise that is completely independent of any human correlation.

phenomena and *noumena*, the latter existing independently of us yet, for Kant, absolutely removed from us. Cage is a Kantian and the Speculative Materialist would want to challenge this thought. The Speculative Materialist could simply say that it is not at all contradictory to imagine or postulate a time in the far future where no particles vibrated and energy was at its lowest state, so no processing could occur, and that would be a *de facto* silence. Similar states that are more trivial also exist, for instance at absolute zero.[17] Therefore, there can be silences, but we can never perceive these because we cannot perceive non-perception. We can think noise outside of perception, outside of a correlation. It exists in any system as an unwanted possibility, which is why noise *qua* noise is always unwanted even in the arts: it destroys the correlation, or, if used at all, it is used to achieve some destructive act in a very careful way. Noise *qua* noise is more than a dangerous supplement, for in any correlationist idea of art noise is destructive and fatal unless a limit is applied. The application of a limit renders noise as a token or symbol, for instance of nihilism or anti-art. However noise in-itself is more radical as it effaces the possibility of any symbolism at all (just as the noise in our simple computer system above crashes the system). Noise *qua* noise is not the elephant in the room but the tiger, and being free in the room and not in some cage (noise as …), it will devour everybody and everything. What is unwanted in music? What is unwanted is total silence, total chaos, or timelessness (time before and after human life). What is required? Human musicians and audiences, an atmosphere of 78.09% nitrogen, 20.95% oxygen, 0.93% argon, 0.039% carbon dioxide, and vibrations within this medium in a frequency range of 12 Hz to 20 kHz and of durations greater than about a second and less than a human lifetime. What this correlation does is make music, as it exists now, an incredibly small fraction of the known universe. There is in principle nothing

---

17 I can modestly lay claim to 65,536 silences in digital PCM data as stored in computer systems and on compact disks. Any set of data that results in a continuous D.C. offset is, in itself, "silent." See *Jliat – Still Life #5: 6 Types Of Silence* edition xi, released in 2000. A 10 second compilation of all 65,536 possible silences on audio CD can also be downloaded from http://www.jliat.com/silence/.

wrong with this, unless any claims are made that this music represents the Real. In the known universe, music as such can only picture the Real as a gross distortion, not withstanding other possible universes and infinities.

Two simple illustrations.

Two American tourists visit Europe and, as they pass from country to country, stay in Holiday Inn hotels where the curtains remain closed and they watch CNN. They return and, though they have visited Europe and can describe it to their fellow Americans, and though we might think they have acted strangely if not stupidly, they have done nothing wrong. They might have noticed the pasta was better in a place called "Italy"—which would be a metaphor for certain desires to use noise in music—but we would not consider that they have experienced or even attempted to experience Europe. A counter argument might be that, as it is impossible to fully experience Europe (or anything), then the action was justified. This would certainly curtail any science, as well as much previous artistic practice, and might be the cost of rejecting or "caging" noise in music. An encounter with Europe would involve an encounter with foreign languages that would be an encounter with noise.

The above is a fiction. A second example is audio CDs. They represent a totalizable set of objects, a set "for humans"—if they were to hear not just that which is familiar—that would consist mostly of unrecognizable noise.[18]

What the loss of the human correlation as music would produce is impossible to define. It is possibly infinite or as large as the known universe and has a temporality of at least trillions of years. How it would be produced

---

18 "All possible CDs" is a thought experiment. An audio CD stores music by patterns of bits; each audio sample is 16 bits, and each second of sound has 44,100 samples, so 16 x 44,100 gives us a second of sound. Multiply by two for stereo, and then by 60 for a minute, then by 74 for the old specification of the maximum duration in minutes of an audio CD. (The fact you can get longer and different formats is for my purpose irrelevant here.) Multiply 16 x 44,100 x 2 x 74 and we get 6,265,728,000 bits. What follows is that, in this CD format, there are $2^{6265728000}$ possible CDs, and no more. Much of this universe would appear for us as noise, yet "noise" would be more representative of its reality.

and perceived or known is likewise not limited. The alternative I can predict with certainty: if noise is made into music it will become sound, vibrations in air of given frequencies and given length made for humans and in the main by humans.[19]

From the brief description of computation we can use a definition to identify music trivially—"A thing is identical to itself if its negation added to it leaves nothing." However, also from computer science, the definition of noise precludes its definition, as "noise causes the program to fail." Computer science defines noise as essentially destructive: coherence, structure and meaning are lost, and with noise so too is the possibility of identification, limit, boundaries, and rules. This may not be wished for, wanted or liked, but it represents the Real in non-correlationist terms. Noise is the Real.

Size Matters

One of the impetuses behind the current philosophical thinking, which is associated with the arts in the form of Speculative Realism and Object Oriented Ontology, is the critique in the sciences of an egocentric universe (one more step from a logocentric, and phallo-logocentric universe of the previous philosophemes). Meillassoux and Brassier note that science has revealed a universe far larger than humanity, one 300 billion years old with

---

19 This includes not only music with animals as sources—for instance, whale song mediation discs—but also objects such as computer generated jazz (e.g., http://www.youtube.com/watch?v=P-Sjgn78rgw) or generative music in general. For example, "Tiklbox composes an endless stream of high-quality beautiful ambient music ideal for relaxation, reflection, meditation & quiet contemplation" (Intermorphic: play with ideas, accessed February 28, 2013, http://www.intermorphic.com/). Or, similarly, "Scape makes music that thinks for itself. From Brian Eno and Peter Chilvers, creators of Bloom, Scape is a new form of album which offers users deep access to its musical elements. These can be endlessly recombined to behave intelligently: reacting to each other, changing mood together, making new sonic spaces. Can machines create original music? Scape is our answer to that question: it employs some of the sounds, processes and compositional rules that we have been using for many years and applies them in fresh combinations, to create new music. Scape makes music that thinks for itself" (Brian Eno and Peter Chilvers, Generative Music, accessed March 1, 2013, http://www.generativemusic.com/).

a future of trillions of years.[20] The contemporary age has, in many spheres, expanded beyond human perception if not comprehension. Trivially, in music technology the 45-rpm record dictated the length of popular songs, the LP the "concept" album. Such domestic realities also played a part in structuring classical music through the length of operas or concertos, similar to the reasons Dutch paintings of the 17th and 18th centuries were made small enough to fit into the houses of the Dutch bourgeoisie, unlike the majestic canvasses of the Louis XIV epoch in France or the massive abstract expressionist paintings of the large lobbies of Manhattan office buildings in the last century.[21] However, just as the trillion escapes us, so does the present-day storage capacity of the media around us. It is not unusual to have in a home terabytes of storage, or on these devices music, which, if it has not already, will soon assume a temporality greater than that of any potential listener's life expectancy. The same could be said of other media, text, or movies, but it is the ubiquity of the mobile phone and musical playlists that offers the challenge: no one can listen to it all. And if it escapes our listening, it breaks the correlationist grip in a practical demonstration of a real that is bigger than ourselves, bigger than human perception. We already know that sound shares characteristics with a larger electromagnetic spectrum, and over 100 years ago synesthesia was an important influence in the development of a modern art that envisaged a unification of meaning across forms, but

---

20 Trillion is the new million! The recent economic crisis has exposed this inhuman number. It is possible to live long enough to count to a million, even a billion, but not a trillion.

21 "Sony had initially preferred a smaller [audio CD standard] diameter, but soon after the beginning of the collaboration started to argue vehemently for a diameter of 120mm. Sony's argument was simple and compelling: to maximize the consumer appeal of a switch to the new technology, any major piece of music needed to fit on a single CD ... Beethoven's Ninth Symphony was quickly identified as the point of reference—according to some accounts, it was the favorite piece of Sony vice-president Norio Ohga's wife. And thorough research identified the 1951 recording by the orchestra of the Bayreuther Festspiele under Wilhelm Furtwängler, at seventy-four minutes, as the slowest performance of the Ninth Symphony on record. And so, according to the official history, Sony and Philips top executives agreed in their May 1980 meeting that 'a diameter of 12 centimeters was required for this playing time.'" Tim Büthe and Walter Mattli, *The New Global Rulers: The Privatization of Regulation in the World Economy* (Princeton: Princeton University Press, 2011), 46.

UN-SOUNDING MUSIC; NOISE IS NOT SOUND

this was always "for us," it never escaped us. And when science did escape our common sense and human world, it left a kind of alienation in being as expressed in the arts,[22] or worse a return to a fundamentalism. So a symphony orchestra appears to operate by some laws that are more *sharia* than scientific.

There are works that challenge human temporality in listening. Amongst others the artist Mattin's recent work[23] comprises more than 40 hours of MP3 recordings, and there are and have been numerous practices that challenge accepted conventions of musical forms and durations.[24] However, these works in the main remain correlational in an explicit or implicit "human" correlation, by which they work as art. Temporality normally corrals music into what is "listenable" and regards anything outside this as outside of a possible listening experience, as data that can provide us with no meaning, as noise. Working with temporality in music as a "real" timescale, not merely human timescales, would require a radicalization of musical form even if it were still limited to sound. We might have to abandon the correlationism of listening, playing, and making for other methods. These are now available.[25]

---

22 See the influential book *The Two Cultures* by C.P. Snow (London: Cambridge University Press, 2001).

23 *IMPROKUP! Improvisation as squatting and living together* (DVD and booklet, w.m.o/r 38, Stockholm, 2012)

24 For example, *ASLSP* (As SLow aS Possible) by John Cage, a performance of which at St. Burchardi church in Halberstadt, Germany, is projected to last 639 years; *Longplayer* by Jem Finer, which is designed to last for one thousand years; or the prolific output of noise artists such as Merzbow. There are also examples of "non-sonic music" in the work of Yoko Ono, Karlheinz Stockhausen, and recently Seth Kim-Cohen and numerous others. However, the idea of sound, human temporality, and music in some form of correlation still predominates and limits the art form.

25 MP3 is a method of compressing sound files. It is "lossy" in that parts of the original sound are lost during compression. What makes working with the MP3 format special is not this feature, but instead that, unlike other formats (WAV, MP4, etc.), MP3 files can be spliced together as larger entities. The ability to combine these in a quick and simple way makes the creation of large structures very simple, almost like the process of DNA synthesis. At its most simple, if you have a folder containing some MP3 files, in Microsoft Windows using the DOS Copy command it is possible to concatenate these into one large and playable file. To do this, open the MS-DOS shell, and in the MP3 folder type "Copy /b *.MP3 big.MP3" (without quotes). Big.MP3 will now contain all the other MP3s in one playable file. It is then simple to create tools with which it is possible to generate and manipulate sound files of lengths impracticable to work with in real time (real time here being real time for humans); see, for example, http://www.jliat.com/1tb/.

It might be that such realities can be experienced only conceptually and imaginatively, only in the imagination that such territories can be traversed. Today I can hold in the palm of my hand sufficient storage capacity for more than seven years of music.[26] Ten of these devices would store an un-listenable quantity—yet one that is by all means an objective reality—and the data that exists in data farms far exceeds any single human experience.[27]

Do we regard this externality as unreal for us, as not sensible, and in its incomprehensibility as noise, or can we find other methodologies for the appreciation of spaces and sizes that are greater than the human? Science has already done this, and the philosophies above attempt the same. The philosopher Laruelle also has a "Real" that is seemingly limited to the human, yet employs a generic matrix that is infinite and incomprehensible.[28] Meillassoux's contingent hyper-chaos, which offers a potential exo-human infinity in his subsequent work, reverts this back to the human scale of possibilities as the source of some future grieving deity;[29] however, the hyper-

---

26 See, e.g., http://www.seagate.com/gb/en/internal-hard-drives/desktop-hard-drives/desktop-hdd/, accessed February 23, 2013.

27 The world's total storage capacity is obviously large and expanding, and difficult to estimate, but as of 2011 one estimate gives 295 exabytes. (1,000,000 terabytes = 1 exabyte.) Therefore, that is, in 2011, 295 million years of potential sound in MP3 format. If this all seems a little unrealistic or crazy, it could be because it is, or it could be that we somehow cannot or do not want to comprehend the Real, or what Meillassoux calls The Great Outdoors. If art is to get real, it has to work with these scales, because they are there (Lucas Mearian, "Scientists calculate total data stored to date: 295+ exabytes," *Computer World*, February 14, 2001, accessed February 23, 2013, http://www.computerworld.com/s/article/9209158/Scientists_calculate_total_data_stored_to_date_295_exabytes).

28 Laruelle's work with Non-Philosophy offers an alternative approach to radicalizing music, for instance in the idea of "small thoughts, everywhere and with every individual" (John Mullarkey and Paul Smith, *Laruelle and Non-Philosophy* (Edinburgh: Edinburgh University Press, 2012), 1). Briefly, the "Non" here is not a negation but similar to the Non in Non-Euclidean geometry, which "opened up" mathematics. Jarrod Fowler has developed from this his concept of Non-Musicology (see http://www.nonmusicology.com/). "'Non-Euclidean' became a by word for non-absolute knowledge … Even the concept of truth was not absolute … Mathematics was open-ended, uncompletable, infinite …" (John Barrow, *The Book of Nothing* (London: Jonathan Cape, 2000), 157). Music once had a mathematical (Pythagorean) ontology; if this were regained it could likewise become an infinite object.

29 The necessity of contingency (Meillassoux's proposal in *After Finitude*) is employed as an ethical resource proposing a future deity who can effectively (bring justice) mourn victims of cruelty who have died and not been mourned. Quentin Meillassoux, "Spectral dilemma," *Collapse* vol. IV: Concept Horror, Robin Mackay, ed., December 2012.

chaos generated by the necessity of contingency is not limited to or for the human.[30] Speculative Materialist philosophies might, in the allure of non-totalizable objects, appear as philosophemes of the Romantic poets, as if the incomprehensibility of these objects were not available to the philosophic mind still bound by a correlationism, or nihilism of life as an existential being. The problematic for philosophers and artists, including musicians, seems to lie in their gaze,[31] a problematic of Petrarch's mountain[32] in looking in the wrong direction into the nature of the subject/object relationship and not into the reality of the object. However, it is now possible to make works of sound that might be considered music even if they are un-listenable not only in practice but also in principle. A simple reaction is to reject this and go back to the cloisters of a medieval universe of music where man is, and always is, at its center. But this is now only an illusion. If noise *qua* noise does come into music as anything more than another trope, it exposes the Real, expanding music beyond the human; how we do this, understand it, appreciate it, or disseminate it is already given by technology.

For Want of a Conclusion

Through the advent of a cybernetics of sound—which generalizes, democratizes, and personalizes music in MP3 playlists on mobile phones, hard drives, and the internet—music, more than any other medium, has direct access to hyper-objects, hyper-chaos, territories finitely and possibly infinitely bigger than the human, which is noise as un-sounding music.

---

30 "Our absolute, in effect, is nothing other than an extreme form of chaos, a *hyper-Chaos*, for which nothing is or would seem to be impossible, not even the unthinkable.... [F]ar from guaranteeing order, it guarantees only the possible destruction of every order," Meillassoux, *After Finitude* , 64.

31 In this case non-Lacanian.

32 On April 26, 1336, Petrarch climbed to the top of Mont Ventoux from where he could see the Alps. The act is often used as a metaphor for the turning point in which medieval thought is re-directed outward to the real world of nature as opposed to the cloistered inner world of the "Dark Ages," a term of which Petrarch is credited with first use.

However, strong forces counter the ideas expressed above of removing the perceptual core of music's ontology. These ideas can be dismissed as fringe phenomena, as peculiar and invalid ideas because they fail to account for the general consensus that music is predicated on sounds that are heard, that the "real" world of music is listening, that the idea of a non-correlationist Real is contradicted by the reality of listening in the "real" world of music. However, this is not the case. The quantities of data in volumes located on YouTube, Bandcamp, SoundCloud, and elsewhere, though created almost certainly with the intention to be heard, cannot in fact be heard. Human perceptual systems can no longer cope with the scale of these quantities. Any attempt at perception will be fractional, distorted, and always incomplete. Overwhelming the perceptual system effaces communication. This has already occurred. And once quantity overwhelms the system, qualitative judgments are no longer possible.[33] And this has also occurred. Within the world of music, mediators of value and taste have changed from the few to the many, and quantity effaces any qualitative ability. In general, everything published on the web is "Awesome!" Obviously this is not immediately good news for anyone who thinks music has a value, or wants to produce value from it. It is a problem for manufactures and retailers of music as product, for reviewers of music and arbiters of taste, and perhaps even for musicological study and evaluation in academic institutions.[34] What has happened to music? Is it now beyond good and evil?

> And do you know what "the world" is to me? Shall I show it to you in my mirror? This world: a monster of energy, without beginning, without end…[35]

---

[33] This use of noise to overwhelm a system is found in nature and in the military. Herding animals confuses the predator by overwhelming its perceptual ability to make a judgment.

[34] This might also represent a problem for the "artist" as some lone Nietzschean *Übermensch* or Romantic genius, a problem for the artist as anything other than yet another of the numberless pop wannabes?

[35] Frederick Nietzsche, *Will to Power* (New York: Vintage Books, 1967), 549.

Alice Kemp (Germseed)

What is noise (music) to you?

I'll keep this personal and not wander into histories or hair-splitting genre definitions. Noise is a living entity, pure expression, force, simple and complex. It is a system of diametrical differences—it can be a cycle of giving and receiving, an exchange of power. It is chaos racket or absolute laser precision. It is extended drone and fragmentary bursts. It can deal in specifics or unknowns. It can be both irritant and balm, purgative, cleansing, medicinal. A source of both exhilaration and panic, fight and/or flight, control and release. It can be what you need to escape from, whilst being just exactly what you need. It can act as social cohesion, compelling, repellent, tribal, or hermitic. It can be deftly brutal or subtlety incarnate. Absurd, perfect slapstick, funny haha, deadpan serious. It can contain portals that allow access to altered states. Crucially, silence can be noise (and again, *vice versa*).

Why do you make it?

Seduced by soundworlds since my early years, I work with these ideas because of their very existence.

Noise and the Voice: exploring the thresholds of vocal transgression

Aaron Cassidy

The voice can say much more and, when it speaks the body, cannot help but speak of other things.
—Douglas Kahn[1]

Noise in vocal artistic practice poses a fascinating problem of definitions. Establishing the thresholds of vocal noise is surprisingly difficult, as the conventional boundaries of noise are not of much use when it comes to the voice. From a spectral, acoustical, Helmholtzian standpoint, almost all vocalization is, to greater and lesser degrees, noise. Unvoiced, unpitched consonants are, by definition, noise, and most voiced consonants include significant noise components. Indeed, even the purest sung or spoken vowels contain aperiodic, non-harmonic components: "the flow of air through the oscillating narrow slit between the vocal folds creates the possibility for turbulent noise, similar to the way we form consonants such as 's' by creating a narrow place in the vocal tract and forcing air through it."[2] Signal processing notions of noise are similarly unhelpful in establishing the boundaries of vocal noise practice, as they rely principally on intentionality and meaning, on some presumption of an attempt towards communicative utterance. While this approach is potentially useful from a linguistic standpoint, for the work under discussion here, which is principally artistic rather than

---

1 Douglas Kahn, *Noise Water Meat* (Cambridge: MIT Press, 1999), 291.

2 Perry R. Cook, "Pitch, Periodicity, and Noise in the Voice," in *Music, Cognition, and Computerized Sound: an introduction to psychoacoustics,* ed. Perry R. Cook (Cambridge: MIT Press, 1999), 202. Cook goes on later to say, "there is some noise that cannot be controlled by the singer and, as in vibrato, there is always some amount of noise no matter what the singer does to eliminate it" (204).

principally textually communicative, the signal itself may well be noise, or at least intentionally noisy, and the notion of "irrelevant or unwanted data" seems out of place. Simon Reynolds provides a third possibility, fascinating but equally unhelpful in establishing the boundaries of noise, proposing that "noise occurs in moments, tiny breakages and stresses dispersed all over the surface of music, all kinds of music." That is, vocal noise appears in the microscopic cracks in the voice, in the way that vocalists use their voices "for the gratuitous voluptuousness of the utterance itself."[3]

This chapter aims to examine vocal artistic noise practice as a performed act of transgression. It explores questions of otherness, of uniqueness and individuality, of revelation of the natural and the performance of the unnatural, and of noise as a decentering, deterritorializing act. And it investigates why we find vocalized noise often so deeply unsettling, strange, or disturbing that we instinctively laugh out of embarrassment or discomfort, while similar instrumental or electronic sounds rarely carry the same baggage.

Despite the proliferation of writing about noise and noise music in recent years, there is surprisingly little written about noise and the voice. It is particularly surprising because the voice has played an important role in theories of noise for the last one hundred years. Consider Luigi Russolo's classifications of the six families of noise in *The Art of Noises* from 1913. In all but the fifth category ("Noises obtained by beating on metals, woods, skins, stones, pottery, etc."), the voice is present throughout, at least implicitly: roars, hissing roars (type 1), whistling, hissing, puffing (type 2), whispers, murmurs, mumbling, muttering, gurgling (type 3), screeching, humming, buzzing (type 4), shouts, screams, shrieks, wails, hoots, howls, death rattles, sobs (type 6). All are clearly vocal, despite the fact that Russolo's manifesto was principally a call to arms for mechanical, instrumental, and electronic music

---

3   Simon Reynolds, "Noise," in *Audio Culture*, ed. Christoph Cox and Daniel Warner (New York: Continuum, 2004), 58.

making.[4] Indeed, though Russolo's categories use the voice as an analogy or a descriptive reference, most noise practice since his time has principally been music for machines of one sort or another.

To that end, this chapter focuses specifically on the unaccompanied, unadorned voice. There are of course numerous examples of musicians and artists who use their voices as part of a noise practice that includes electronic modification and distortion, from the early work of Henri Chopin to Trevor Wishart to more recent practitioners such as Maja Solveig Kjelstrup Ratkje[5] or indeed much of the vocalization in the "noise music" genre. I am intentionally avoiding discussion of that kind of approach here primarily because it seems necessary, in this context, to direct the discussion not toward the use of the voice in noise but instead toward noise in the voice itself. That is, this chapter is not about noise practice that incorporates the voice; it is about the noise of the voice used in artistic practice and establishing what the thresholds of vocal noise might be. (The only extension of the voice that appears in the repertoire below is the microphone.) Despite that limitation, the scope is fairly broad, including examples from free improvisation, grindcore, sound poetry, performance art, notated music, and theater.

Vocal Noise, Otherness, and Limits of the Body

There is a moment in Antonin Artaud's *Les malades et les médecins* (1946) that I find particularly fascinating.[6] The opening few minutes are typical

---

4 Luigi Russolo, "The Art of Noises," excerpted in *Audio Culture*, 13.

5 As Marie Thompson puts it, "Ratkje cracks open the voice to find a mouth full of wires." Marie Thompson, "Music for Cyborgs: the affect and ethics of noise music," in *Reverberations: the philosophy, aesthetics and politics of noise*, ed. Michael Goddard, Benjamin Halligan, and Paul Hegarty (London: Continuum, 2012), 217.

6 Antonin Artaud, *Deux Émissions De René Farabet* (France: André Dimanche Éditeur, 1995). The audio is also available on YouTube (http://www.youtube.com/watch?v=RaQ-KG4cjrI) and ubuweb (http://www.ubu.com/sound/artaud.html). An unattributed translation of the text can be found here: http://www.falseart.com/antonin-artaud-les-malades-et-les-medecins/.

of Artaud's approach to vocalized text: odd, singsong, hesitant, faltering, mannered, with a consistent grit and brittleness in the voice, and occasional higher pitched, childlike squeaks and falsetto blips. Just after the three-minute mark, the text, which had been bizarre and mangled but was nonetheless text, is briefly replaced by a shocking, gripping, horrifying series of indecipherable, chicken-like squawks and yelps. It takes only a few seconds, a microscopic percentage of the five-minute overall duration, but it is astonishing. A line is immediately crossed from text that is *about*, however loose or strange, to a state of otherness that is both revelatory and distancing.[7] For me, the fascination lies in the immediacy of this leap into otherness. The voice continually pushes against the threshold, but, as though pushing against a spring that suddenly collapses, the moment the threshold is crossed a seemingly insurmountable gap into otherness is formed. It is a transgressive moment that opens a fissure that cannot be closed.

This sense of otherness is a common thread in vocal noise. The spaces that it opens are strange, funny, bewildering, embarrassing, shocking, and enthralling; they seem to draw us in as listeners to a peculiarly personal place, which can be both intimate and inviting or disturbing and repelling. There is a vulnerability to this kind of vocalization that is quite distinct from similar sounds made through non-vocal means.

> The voice comes to us as an expressive signal announcing the presence of a body and an individual. ... The voice is the very core of an ontology that balances presence and absence, life and death, upon an unsteady and transformative axis. The voice comes to signify through a slippery and unforgettable semantics the movements of consciousness, desire, presence, while also riveting language with bodily materiality.[8]

---

[7] Douglas Kahn describes an account of Artaud's screaming during a lecture on "The Theatre and the Plague." "Afterward among friends, Artaud protested, 'They always want to hear *about*; they want to hear an objective conference on "The Theatre and the Plague," and I want to give them the experience itself, the plague itself, so they will be terrified, and awaken.'" *Noise Water Meat*, 349 (emphasis added).

[8] Brandon LaBelle, "Raw Orality: Sound Poetry and Live Bodies," *Voice: Vocal Aesthetics in Digital Arts and Media*, ed. Norie Neumark, Ross Gibson, and Theo van Leeuwen (Cambridge: MIT Press, 2010), 149.

This link between the voice and the body is made particularly evident in the extraordinary work of the British vocalist Phil Minton. A leading vocal improviser, Minton has generated a repertoire of nearly inexplicable extended techniques that push at the limits of what is possible with the voice. I have seen numerous concert reviews with lists of verbs attempting to describe what Minton does, but none come close to explaining what the sounds actually are, predominately because those descriptors usually rely on references to a known collection of sounds (belching, growling, Donald Duck) that dramatically minimize and reduce the actual palette of vocalizations in his work, which are quite a bit stranger.

I am struck by the disparity between how I listen to Minton's work when I can see him, either in person or on video, and when I cannot.[9] It reveals a curious gap between the pure physiology of the voice and the physiology of the voiced body, that is, the gap between the actual sound production mechanism of larynx, breath, tongue, and mouth versus "the voice" as the representation of human physicality and expression. Sonically, we hear the voice, but we hear such an otherworldly voice that it is often strangely disembodied. The further away the vocalizations get from sung or spoken vocal production, the more this phenomenon is magnified, to the point that all that is left is "grain," a pure, vocal materiality.[10] But in performance, these sounds are coupled to Minton's twisted, contorted bodily motions and to a most extraordinary set of facial expressions, which often look similar to the smeared and distorted faces of a Francis Bacon painting: as Deleuze puts it, "the forces of deformation, which seize the Figure's body and head, and become visible whenever the head shakes off its face."[11] Though he typically

---

[9] For example, http://www.youtube.com/watch?v=wCS4vUym0_8 or http://www.youtube.com watch?v=g2OLYsOlHNA.

[10] Roland Barthes, "The Grain of the Voice," *Image-Music-Text* (New York: Hill and Wang, 1978), 182, 188.

[11] Gilles Deleuze, *Francis Bacon: The Logic of Sensation*, trans. Daniel W. Smith (Minneapolis: University of Minnesota Press, 2004), 53.

sings seated, generally with his hands comfortably on his knees, his singing is bodily, the voice only the final extension of a deep physicality.

> The human voice readily becomes its Other through vocalization. Conversely, the face retains its visible base onto which are mapped the mannered means for shaping itself into facial variations. The pulled face, the disfigured face, will always be marked by its human physiology. The voice—through the multiplicity of its invisibility—will always locate itself in adverse directions beyond the origin of its generation and production. Like the thrown voice coming from somewhere else, the voice sings, speaks, screams, and states its residence in Otherness.[12]

Minton's work also draws attention to a phenomenon common to vocal noise more generally. We hear struggle in his singing. In works such as "Wreath" or "Blasphemy," from the collection *A Doughnut in Both Hands*, we hear the raw physicality and extremity of some of Minton's more severe vocal practice.[13] Both works have moments in which vocalization dissolves into actual gagging and coughing as Minton gasps for breath or chokes on his own spit and apparently, in the case of "Wreath," vomit. In "A Good Song" or "Psalm of Evolution 1," from the same disc, the form of the improvisation seems driven at least partially by the amount of time Minton can maintain any one sound—we hear him "reset" periodically, the limits of the body pushing back against his insistence towards sounds at the threshold of tension, pressure, and breath. We can hear that he is *forced* to stop, to swallow, to catch his breath, to recover at least temporarily from the strain placed on the vocal mechanism.

One of the critical thresholds of otherness and transgression in noise vocalization is the sense of physical tension, discomfort, constriction, and even pain present in the voice, and, more significantly, the way in which

---

12 Philip Brophy, "Vocalizing the Posthuman," in Neumark, Gibson, and Van Leeuwen, eds., *Voice*, 361.

13 Phil Minton, *A Doughnut in Both Hands: Solo Singing 1975–1981* (Emanem 4025, 1998).

those sounds are elongated, magnified, and intensified. It is their extremity and their relentlessness that carries them into the realm of noise. It creates an initial empathy with the listener, an awareness of risk, and then pushes that threshold of risk up to and occasionally beyond the limits of the body. Georges Aperghis's *Récitation 14*, for example, asks the singer to breathe in, hold the breath for as long as possible, and then speak a repeating, looping rhythmic cell with cross-cutting French and English texts, with a tempo that slows consistently from a quarter note pulse of 120 to 40 BPM over the course of 18 repetitions, before the whole process begins again with a new breath.[14] The final cell of the 18 ends with four eighth-note rests—an excruciating three-second pause, all available breath by this point well since expelled and the brain starved of oxygen—so the repetition begins not with the careful, deep breath that starts the piece but instead with a gasping, fearful breath of self-preservation before, again, pausing as long as possible before beginning the rhythmic looping text. The singer is asked to repeat the process "ad libitum," though few singers seem to manage it more than twice. The process dramatically affects the kind of vocal production employed. The singer, carefully rationing breath, typically uses a halting, constricted sound, rather than the more playful, balletic sung and spoken text common to the rest of the *Récitations* collection. The pressure and restriction is palpable, and, as is the case with the Minton examples above, establishes an immediate sense of empathy with the listener and an awareness of the very real physical vulnerability present in the work.[15]

The work of the performance artists Ulay and Abramović goes even further. Much of their work during their twelve-year partnership in the 1970s and 80s included investigations and interrogations of bodily limits,

---

14   Georges Aperghis, *Récitations* pour voix seule (Paris: Editions Salabert, 1982).

15   I am thankful to the soprano Peyee Chen (who, incidentally, managed three repetitions) for her very helpful insight into the experience of performing the piece.

often invoking extreme bodily fatigue, pain, and risk of death.[16] The piece *AAA AAA* (1978) provides an ideal example of the thresholds of noise in the voice:

> We are facing each other both producing a continuous vocal sound. We slowly build up the tension, our faces coming closer together until we are screaming into each other's open mouths.[17]

As with Artaud's *Les malades et les médecins*, the threshold here is immediately apparent and helps to establish how aspects of both otherness and physical limitations contribute to an awareness of noise in the voice. What begins as a controlled, almost sung yell moves closer and closer to the threshold of noise. Within a few minutes, the strain placed on the voice through the yelling starts to introduce small quivers and yodels, the voice increasingly out of the control of the vocalist. The end of each yell begins to reveal a raspy, abrasive roughness in the voice, the internal discomfort and pain in the throat inching closer to the aural surface. After several minutes of continuous yelling, Ulay's voice is the first to break—he coughs instinctively and involuntarily, and even after he regains his vocal composure we see him swallow noticeably after the end of the next yell in a feeble attempt to soothe the throat with saliva. As the screams intensify, the need to reset and recover intensifies as well, the scene increasingly poignant as the two vocalists move closer and closer together, their noses now almost touching. By the fifth minute, their heads start to shake during the yells, and the time between yells gets longer, the yells shorter and louder, the pain in the voice even more apparent. It is here where the threshold seems to be crossed into a transgressive space of otherness. Perhaps

---

16 See for example *Relation in Time* (1977), in which the two sat back-to-back with their hair tied together for 16 hours, *Rest Energy* (1980), in which the two stand face to face holding a drawn bow and arrow between them, or *Breathing In/Breathing Out* (1977), their open mouths locked together, one breathing out as the other breathes in as they slowly poison each other with carbon dioxide until they eventually collapse, unconscious.

17 Ulay (Uwe Laysiepen) and Marina Abramović, *AAA AAA* (1978), accessed April 11, 2013, http://www.youtube.com/watch?v=iAIfLnQ26JY. The text is Abramović's voice-over introduction.

it is simply the place at which almost anyone else would stop. And perhaps it is also this moment that reveals the otherness of the performer: our empathy, our befuddlement, our discomfort, and our instinctual embarrassment each contribute to the them-ness of the vocalists. The transgressive act reveals their humanity, but it also reinforces that they are not us.

The grit and vocal fatigue continue to intensify, the yells become increasingly bodily. Nine minutes in, Ulay gives up, unable to yell anymore, while Abramović continues for another fifteen seconds or so on her own. We see and hear the moment when the voice fails, where it is simply no longer possible to continue to vocalize.

> For Artaud the idea of *asphyxiation*, which he employed throughout his career, ran across breath, air, and social space as a means to mediate between life and death corporeality and the suffocating effects of society. Indeed, asphyxiation and the scream formed the parameters of the complex of this exchange—one a withholding of breath/air/space, the other its ultimate activation, both determined by total purpose. ... What was once the actual body has been left intact and rendered fantasmic, reduced to a double that exports a few organs into the functioning, vacant body when called for. These organs no longer retain their normal functions but have been reduced to serving only as signposts to locate the movements of breath and void in the vacated, actual body.[18]

It is a very *real* noise, an authentic noise. There is a social space that is transgressed (verifiable even with a quick glance at the hundreds of comments on the YouTube video of *AAA AAA*) as well as a transgression of the physical

---

18 Kahn, *Noise Water Meat*, 351–52.

limits of the voice.[19] Its noise confronts us with otherness in large part through the uncontrollable, instinctual, bodily reaction and resistance, its separation from the controlled space of the sung and spoken voice.

> The articulate singing voice that projects an image of the conscious human—capable of language and social expression—is the voice masquerading as a living force. Its recourse to language remits it to symbolically celebrate its distance from death. ... But that same remit blocks the articulate singing voice from confronting voice at its elemental, its guttural, its expulsive. Posthuman vocalization thus refers precisely to how the voice returns to breath.[20]

Vocal noise artists draw on this sense of otherness, and indeed on the identification of physical tension and pain, amplifying and reveling in the noise of the broken voice. The Canadian group Tunnel Canary, one of the earliest practitioners of noise rock, exemplifies this practice:

> I first got the idea for the style of the vocals at an early Vancouver punk gig when a bunch of punk girls were fooling around with a stage mic between sets. One of them (it may have been Mary Jo, bass player for the Modernettes) let out a loud scream. I thought this was the most interesting and powerful event of the evening. *Too bad, I thought, that it couldn't have gone on longer.* ... I decided to experiment with this idea and asked Ebra (then my girlfriend) if she could scream for an extended period of time. She tried and was incredible.

---

19 As Norie Neumark writes, "to allow [the voice] to rise in a scream is to open the body to be read as hysterical, infantile, out of control, uncivilized. The scream is the sort of improper vocal gesture that proper bourgeois subjects learn not to emit." Norie Neumark, "Introduction: The Paradox of Voice," in Neumark, Gibson, and Van Leeuwen, eds., *Voice*, xxviii.

20 Brophy, "Vocalizing the Posthuman," 362.

A screaming, angry female (Earth Mother taking her revenge) voice over radical, improvised electronic sounds—thus Tunnel Canary was born.[21]

The transgression exists largely in the artistic practice of funneling into a more or less natural noise—in this case a scream, in the case of Phil Minton a gravelly, squeaky, broken, or multiplied voice—and magnifying the noise, exploding it out to the surface and extending it to a breaking point.

Vocal Noise as Nomadic Space

Language has a territorializing function. It establishes boundaries and relationships, hierarchies, and connections through short- and long-term memory. Noise on the other hand is a fundamentally deterritorializing phenomenon, what Deleuze and Guattari might call a "local space of pure connection."[22] Indeed its noisiness lies principally in its destabilizing, its upending of communicative norms and hierarchies.

Here, there are no longer any forms or developments of forms; nor are there subjects or the formation of subjects. There is no structure, any more than there is genesis. There are only relations of movement and rest, speed and slowness between unformed elements, or at least between elements that are relatively unformed, molecules and particles of all kinds. ... We call this plane, which knows only longitudes and latitudes, speeds and haecceities, the plane of consistency or composition (as opposed to

---

21   Nathan Holiday, "Tunnel Canary," Museum of Canadian Music, accessed April 12, 2013, http://www.mocm.ca/Music/Artist.aspx?ArtistId=104558&RoomId=57, emphasis added. Holiday's account of the formation of the band also includes the following amusing anecdote: "I remember a funny incident at one of the first performances where two policeman thought Ebra was either deranged or having a seizure and approached her very cautiously; they were terrified, you could see it in their faces. When they finally asked her what was wrong, and to calm down, she told them that she was singing. They got very mad and told us to get the hell out of here and don't come back."

22   Gilles Deleuze and Félix Guattari, *A Thousand Plateaus* (Minneapolis: University of Minnesota Press, 1987), 493.

the plan(e) of organization or development). It is necessarily a plane of immanence and univocality.[23]

Deleuze and Guattari refer to this phenomenon in many parallel ways—smooth space, haptic space, the plane of consistency, etc.—each emphasizing the local, immediate, and the unbounded. It seems an ideal description of noise more generally and vocal noise specifically.[24]

Sound poetry revels in this decentered ambiguity. The work of The Four Horsemen, a Canadian performance collective active in the 1970s, for example, explores a vocal poetic practice that foregrounds an untethered, scrambled, non-semantic verbal communicative space.[25] It is a communicative art, but an intentionally non-textual communication, in what the group's founder, Steve McCaffery, describes as an "intermedia experience ... generated on the liminal zones of theatre, music and poetry."[26] The group's work reorients vocal communication to a principally bodily rather than textual practice, including yells, screams, grunts, and various physical manipulations of the mouth, as well as unpredictable mutations and distortions of localized phonemic content through elongation and repetition. This approach is similarly seen in the solo work of McCaffery and Paul Dutton (another member of The Four Horsemen). For example in Dutton's "Mercure" or in the collection *Mouth Pieces: Solo Singing*, or in McCaffery's performances of the concrete poem *Carnival*, a common technique is grabbing, stuttering, and repeating a short

---

23 Deleuze and Guattari, *A Thousand Plateaus*, 266.

24 See Paul Hegarty "A Chronic Condition: Noise and Time," in Goddard, Halligan, and Hegarty, eds., *Reverberations*, for a similar discussion about this deterritorializing function of noise with regard to time and form.

25 See for example this excerpt of Ron Mann's 1982 film *Poetry in Motion*, http://www.youtube.com/watch?v=843O0bTVKHQ.

26 Steve McCaffery, "Sound Poetry – A Survey," from *Sound Poetry: A Catalogue*, ed. Steve McCaffery and bpNichol (Toronto: Underwich Editions, 1978). Reproduced at http://www.ubu.com/papers/mccaffery.html, accessed April 19, 2013.

fragment of text or a phonemic subcomponent of a word, or, similarly, taking some small subcomponent of the production of a word or simple vocal sound and extending and stretching it, both revealing its internal, microscopic sonic reality and turning it into something else altogether.[27]

We might identify sound poetry as a cultural arena granting witness to the movements of certain bodies on the way in and out of communicative acts. It seeks to rivet language with new sonorous materiality and in doing so refashions the self's relation to vocality and processes of signification. ... [I]n seeking other relations to speech, it retools the mouth by incorporating an oral calisthenics, concocting conditions for other linguistic acts, literally seeking to bypass the regular movements of orality for new configurations, and turning the mouth into the site of production for other semantics. That is, sound poetry yearns for language by rupturing the very coherence of it.[28]

Gil J Wolman's work is even more extreme in its abandonment of the territorializing features of language, leaving behind even the basic phonemic content of vowels and consonants. Active in the Letterist movement in the 1950s, Wolman developed the concept of the "megapneume" (*mégapneumie*), "a poem of great breath or vigor, a mega-breath, ... a striking form of *physical poetry*."[29] In "La mémoire," perhaps the most absolute example, the content of the megapneume is exclusively the sound of breath: inhaling and exhaling, a quivering and pulsation of breath, modifications of the size and shape of the

---

27 Paul Dutton, "Mercure" (http://www.youtube.com/watch?v=zCaCHyj4ozk) and *Mouth Pieces: Solo Soundsinging* (OHM Editions, OHM/AVTR 021, 2000), also available at http://www.ubu.com/sound/dutton.html. Steve McCaffery "Carnival," live performance from Instal Glasgow, 2009, accessed July 31, 2013, http://www.youtube.com/watch?v=Z5sB_YvvSS4.

28 LaBelle, "Raw Orality," 150.

29 Frédéric Acquaviva, "Wolman in the Open," Museu d'Art Contemporani de Barcelona, accessed April 19, 2013, http://www.macba.cat/PDFs/acquaviva_eng.pdf. Italics in original.

mouth to color and filter the pitch and contour of the breath, a distinction between breath through the nose and breath through the mouth, and changes in the speed and duration of breathing.[30] It is quintessential vocal noise, both in its sonic character and in its deterritorialized, localized communicative space. Its noise is, as Deleuze and Guattari put it, a nomadic line:

> *A line that delimits nothing, that describes no contour*, that no longer goes from one point to another but instead passes between points, that is always declining from the horizontal and the vertical and deviating from the diagonal, that is constantly changing direction, a mutant line of this kind that is without outside or inside, form or background, beginning or end and that is as alive as a continuous variation—such a line is truly an abstract line and describes a smooth space.[31]

Devoid of any lingering linguistic ciphers, our focus is drawn to the breath, the body, the voice. Its materials are a scrambling of codes that force a local-level listening, an ambiguity of intention and relationship. The saxophonist Dror Feiler describes this phenomenon in noise music more generally:

> Our attention funnels into the work's singular moments, and once we realize the "NOISE MUSIC" is not here to fulfill a macro-structural objective, it becomes something that ends in itself. Instead of singular "NOISE" existing for the abstract achievements of the whole, the whole is composed to throw us back onto the horns of the "NOISE." ... Every "NOISE" in the music takes on a specific meaning, and no clear hierarchy exists between them. ... Yet equality does not slip into interchangeability, for each "NOISE" in the music remains painfully particular. ... As soon

---

30 Gil J Wolman, "La mémoire," originally released on Revue OU No.33,1968. Also available at http://www.youtube.com/watch?v=372y4Cc7T08.

31 Deleuze and Guattari, *A Thousand Plateaus*, 297–98.

as we encounter "NOISE MUSIC" we are engaged in a struggle to make some sense of what we hear. Unable to categorize the stimulus within any known musical genre, incapable of interpreting or recognizing sounds, and generally bereft of aesthetic orientation, the work commands our full attention. With our ear tuned and focused to hunt out some structure and reason in the work, micrologics emerge.[32]

Marie Thompson puts it similarly:

Noise music in general may be understood to be, quite literally, the musicalization of indiscernibility. Noise music relies on an exploitation of the grey area of contradiction; it lies on the fault line between music and non-music, wanted and unwanted, the pleasurable and grotesque, and so on, pulling in various, conflicting directions.[33]

This phenomenon takes on a slightly different form in the work of the German vocalist Ute Wassermann, whose chirruping, hooting ululations, glottal trills, and yodels are of a completely different sort than any of the other vocalists mentioned here so far. One of the things I find most fascinating about Wassermann's unique vocal technique is the very instrumental, and sometimes even electronic, implication of the soundworld. Unlike the work of Minton, The Four Horsemen, or Wolman, which relies on an approach to the voice that is pared down and often uncomfortably revelatory, this is a much more stylized, intimate, and seemingly controlled vocal practice. The destabilizing factor here is, at least in part, the result of the disconnect between the sound of Wassermann's fascinating, stunningly individual fluttering and our ability to connect those sounds to the mouth, to the voice. To be fair, most

---

32 Dror Feiler, "Exile as Noise – Noise as Exile," accessed April 19, 2013, http://www.tochnit-aleph.com/drorfeiler/exile.html.

33 Marie Thompson, "Music for cyborgs: the affect and ethics of noise music," in Goddard, Halligan, and Hegarty, eds., *Reverberations*, 215.

of Wassermann's work is quite clearly vocal in its character and identity, but for my ears the moments that are most fascinating, and indeed the moments that seem to move into a realm of noise, are the moments in which the voice "goes beyond itself" into a much less identifiably vocal space.

Wassermann's piece "trill territory" on the solo CD *birdtalking*, for example, begins with a fairly simple timbral trill from the tongue on a single sustained pitch, initially a flipping of the front of the tongue as in the consonant "L." But very quickly—in fact in a matter of only a few seconds—the trill moves further and further back along the tongue into the back of the throat into an increasingly unusual, fascinating, and even befuddling flickering that is somehow less explicitly vocal.[34] Similar processes are repeated frequently throughout the improvisation, with numerous sounds seemingly disconnected from the voice, the soundworld dissolving into the sounds of birds, of machines, of electronics. At the very least, our ability to "reverse engineer" the vocal aspect is seriously compromised in a way that is quite exceptional among vocalists. That is, if Phil Minton's "noise" comes largely from a disconcerting interrogation of the realities of the limits of the voice, Wassermann's "noise" comes from a disconcerting unreality, from a vocal practice that is so unique and specialized that it seems to disconnect itself from our ability to understand its sounds as vocal. To put it another way, its noise comes from the *elimination* of the "grain."[35]

This noise is similarly deterritorializing. Its reinvention of the voice destabilizes and decenters the voice and dissolves the relationship between the voice and the normal referencing boundaries of language, song, or even the mouth on a purely physiological level. Wassermann's voice takes the mouth and tongue—or even in a more general sense the voice-as-body—

---

34 Ute Wasserman, *birdtalking* (NURNICHTNUR, 2006).

35 This apparent separation between sound and the physiological aspects of its production can be seen periodically in Wassermann's improvisation "Ute on the Marshes," in the film by Helen Petts made for the "Voice and the Lens" symposium at the Icon Gallery Birmingham, November 2012, accessed April 22, 2013, http://www.youtube.com/watch?v=aQcRpOh3LCc.

and shifts, distorts, and reinvents the connections between these body parts themselves and between the body and the sounds that might result, creating a transgressive space that destabilizes our ability to connect the voice to sound.

Vocal Noise as the Transgression of Uniqueness

Vocal noise is about the individual body, about embodied otherness, about an individualized practice. But one of the great ironies of vocal noise is that, despite its seemingly personal, expressive qualities, its essential "noisiness" comes from the fact that vocal noise is, at its core, an erasure of uniqueness.

The voice in its normal state is one of the chief identifiers of individuality and personality, of one's particular person-ness.

> What it communicates is precisely the true vital and perceptible uniqueness of the one who emits it. At stake here is not a closed-circuit communication between one's own voice and one's own ears, but rather a communication of one's own uniqueness that is, at the same time, a relation with another unique existent.[36]

Noise erases, or at least sidesteps, this uniqueness. On both a spectral and physiological level, this erasure comes from the fact that the vocal production of noise largely eliminates the resonating cavities that create vocal formants. In essence, it severs the connection between the voice and the body. Simple unvoiced consonants—s, sh, ch, k, t, etc.—require very little of the body for resonance. They are made almost entirely in the mouth, and their spectral content is created almost exclusively from the shape and size of the mouth and tongue. The difference between my "s" and your "s" is minimal. Vowels and voiced consonants, on the other hand, engage the body; their character,

---

36 Adriana Cavarero, *For More Than One Voice: Toward a Philosophy of Vocal Expression*, quoted in Neumark, "Introduction," xix.

weight, and color come not only from the size and shape of the vocal folds, tongue, and mouth, but also from the chest, head, and bone structure of the speaker or singer. It is the uniqueness of these resonating cavities that gives a voice its identifiable connection with its emitter, and the noises of the voice sidestep this resonance and thus this identification.[37]

Vocal noise practice extends and plays on this gap. It lives in and exploits the transgressive "otherness" of what is, in a sense, a "voiceless" space. It is the space of the witness whose voice has been modified to hide his identity, a disembodied voice stripped of its primary identifying features. "Noise routes expression through the impersonal and indulges in a catastrophe of generic gestures."[38] It is of course not only the noise of unvoiced consonants but also the performance of this otherness more generally that distinguishes vocal noise through its intentional, *performed* voicelessness. In the vocal performances of Valeri Scherstjanoi, for example, we have guttural grinding, clicks, hisses, and grumbles—it is a unique and personal artistic practice, but it is not the sounds themselves that are unique. It is instead the personally unidentifiable aspect of his vocalization that is the most definitively *noisy*. This intentional elimination of personality is a transgressive act similar to the art of Cindy Sherman,[39] with its performed erasure of identity, or the work of Franko B, which is deeply personal and vulnerable and yet also an erasure of the personal in its whitewashed elimination of many of the defining features of face and

---

37 This sense of transgression may also have something to do with the evolutionary nature of consonants vs. vowels. Anne Karpf, in *The human voice: the story of a remarkable talent* (London: Bloomsbury, 2006), explains that consonants are evolutionarily much older than vowels, with human speech as we know it only about 100,000 years old as pitched vocalization only emerged through the relatively recent physiological dropping and lengthening of the larynx (52–53, 55). It is purely speculative, but perhaps this elimination of identifiably personal vocalization in noise is also an instinctual connection to a pre-modern, early evolutionary human state?

38 GegenSichKollektiv, "Anti-self: experience-less noise," in Goddard, Halligan, and Hegarty, eds., *Reverberations*, 194.

39 Cindy Sherman, *Cindy Sherman: a play of selves* (Ostfildern: Hatje Cantz, 2007).

body, or the photography of Aziz + Cucher, which erases the defining features of the face (eyes, mouth, and nose) through digitally reconstructed "skin."[40]

Ami Yoshida's improvisations are similar.[41] Though her vocalizations are in many ways extraordinary in their sonic inventiveness—impossibly high shrieks, throat whistles, guttural grinding, piercing wails—they are also vocalizations that disconnect the voice from the body. These sounds are made in a microscopically small physical space in the throat. Like the squeal of air coming out of the pinched opening of a balloon, or like the unpredictable squawk from a clarinet reed, they are the noises not of their resonating chambers but instead of a tiny, compressed, pressurized space, here even smaller and even deeper into the recesses of the throat than the sounds found in Scherstjanoi's work. (This perhaps explains how, as a tall Caucasian male, I can manage to make similar sounds as Yoshida, a short Japanese woman.) Yoshida's performance mannerisms intensify this sense of separation, her eyes often either closed or looking up towards nothing in particular, which seems to draw our attention from the face and the torso inward towards the mouth, neck, tongue, and throat.[42]

This performed, noisy voicelessness is particularly present among vocalists whose work aims to make connections with the occult and with notions of evil. Take, for example, the vocal practice of "grindcore" (or to an even greater degree the subgenre of "pig squeal").[43] This gritty, gravelly vocal style revels in an ambiguous voicelessness. Its (literally) abrasive technique eliminates

---

40 Lea Vergine, *Body Art and Performance: the Body as Language* (Milan: Skira, 2000), 273, 276.

41 See, for example, Ami Yoshida, *Tiger Thrush (Improvised Music from Japan,* IMJ-504, 2003), also available at http://www.youtube.com/watch?v=bF4QszdBvl0.

42 "Power of Making; Ami Yoshida, Finding Voice – the Lingual," accessed July 25, 2013, http://vimeo.com/27164111.

43 There are numerous YouTube genre compilations available. See for example http://www.youtube.com/watch?v=iq_CajFb_5s or http://www.youtube.com/watch?v=k-3maJ3 for grindcore samples, or http://www.youtube.com/watch?v=3FpFkazuUmo for pig squeal.

the controlled, harmonic vibration of the vocal folds and greatly hinders the ability to produce and pronounce words in any conventional sense (indeed, some grindcore foregoes any pretense of lyrics altogether, using the voice exclusively for its noise-making abilities). The technique subverts the natural timbre of the voice, replacing it with an intentionally personless, unidentifiable noise, an observation that seems to be reinforced by the numerous "how to" instructional videos and webpages on producing a "grindcore voice."[44] Rather than an "Otherness," this vocal practice is a "No one-ness," and it is precisely this subversion of the personal individuality of the voice that crosses the threshold of noise.

> Rather than trying to reconcile knowing and feeling, noise can help us to dissociate the commensurability of experience and subjectivity in a sense that exceeds the logic of framing. ... Noise, with its epistemic violence, counters the division between activity and passivity. By making us aware of our inability to decipher it, noise alienates us.[45]

Critically, this approach to noise production is highly intentional and, to a certain extent, even theatrical. The sounds produced are not actual screams, cries, or moans, and the more extreme the sound production employed, the less natural and innate the voice seems to be. That is, the voice is less personal and unique and less directly connected to its bodily creator. Norie Neumark uses the example of staged laughter:

> it's the convulsions of broken voices that remind us how the voice normally coheres, despite its inherent tensions. Laugher, as it breaks into the voice, is an outburst of physiology and culture—both individually and

---

44 See, for example, http://www.wikihow.com/Perform-Grindcore-Vocals, http://www.youtube.com/watch?v=Pq2Nrr8qkkM, or http://www.youtube.com/watch?v=UFtfCm8ogTM. (There are, of course, also parody videos: http://www.youtube.com/watch?v=Y4qSiM6Bgvg.)

45 GegenSichKollektiv, "Anti-self," 194.

infectiously from one body to another. Laughter has both an immediate and mediated relation to voice. Because laughter is voice out of control—released at the edge of voice—controlled laughter, from the individual's forced laugh to the TV laugh track, can feel particularly disturbing.[46]

The "noise" of the voice, in this case, comes from the staged, inauthentic costume of the voice of the Other. It is not the sounds themselves that are noise, as such, but instead the act of taking the tiny breakages of the voice and staging them, elongating them, magnifying them. The discomfort arises through the transgression of making the accidental intentional.[47]

In general, what distinguishes noise in the voice is not so much a specific set of sonic characteristics or descriptors as much as a particular usage of those characteristics. Noise is behavioral, not absolute, and it exists in the thresholds of the voice as an expression of Self. It emerges in the spaces in which vocalization rejects the limits of the self, both physically and psychologically, and in which the articulating function of the voice is subsumed, either through the dissolution of linguistic structures or its timbral, personal individuality.

---

46 Neumark, "Introduction," xxvi–xxvii. One is also reminded of the Masahiro Mori's notion of the "uncanny valley" in robotics and artificial intelligence.

47 Note that this is different from the "self-subversion" discussed by Marie Thompson (quoting Simon Reynolds) in "Music for Cyborgs" (217–18). This is not a subversion of an imagined, invented, straw-man audience.

NOISE IN AND AS MUSIC

Maja Solveig Kjelstrup Ratkje

What is noise (music) to you?

Noise music is liberating, life-celebrating, and demanding. At its best it is multi-dimensional, it is both dark and light, and in its own way soaringly beautiful. The scene is the most welcoming of all genres and there is always a good discussion of aesthetics and tools.

Why do you make it?

Because of all the above! And because I feel that the real noise today is the shallow popular culture, the sell-outs with self-indulgent messages that I can't get around, no matter where I move in civilization. The main reason I make noise is the heartfelt optimism I feel in that expression, as if noise music contains so many more aspects of life than music with an outspoken "message."

NOISE IN AND AS MUSIC

Subtractive Synthesis: noise and digital (un)creativity

Aaron Einbond

> *Invention, it must be humbly admitted, does not consist in creating out of void, but out of chaos.*
> —Mary Shelley[1]

Source/Filter

Subtractive synthesis is a technique of analog and digital electronic sound production that "consists of submitting a spectrally rich wave to a specific type of filtering, thus arriving at the desired tone by eliminating unwanted elements rather than by assembling wanted ones."[2] One common source wave is electronically generated white or pink noise. Early uses included Karlheinz Stockhausen's *Gesang der Jünglinge*, Joji Yuasa's *Projection Esemplastic for White Noise*, and the commercial synthesis of drum sounds.[3] These uses may not normally be termed "noise music," but this noise lies beneath the surface of the history of electroacoustic musics.

This source/filter model is not unique to electronics: it is also a model for acoustic sound production, including the human vocal tract and source/filter coupling in wind instruments.[4] In the example of vocal vowels, the source

---

1 Mary Shelley, introduction to *Frankenstein*, quoted in Jonathan Lethem, "The Ecstasy of Influence: A Plagiarism," in *The Ecstasy of Influence: Nonfictions, Etc.* (New York: Vintage Books, 2011), 98.

2 Jean-Claude Risset and David L. Wessel, "Exploration of Timbre by Analysis and Synthesis," in *The Psychology of Music Second Edition*, ed. Diana Deutsch (San Diego and London: Academic Press, 1999), 128.

3 Jason Anderson, "Slaves to the rhythm: Kanye West is the latest to pay tribute to a classic drum machine," *CBC News*, November 28, 2008, accessed May 6, 2013, http://www.cbc.ca/news/arts/music/story/2008/11/27/f-history-of-the-808.html.

4 Wayne Slawson, *Sound Color* (Berkeley: University of California Press, 1985), 22–28.

is a pitch produced by the vocal chords and filtered by the cavity formed by the throat and mouth, or the noise arising from turbulence to produce a consonant and filtered by the vocal cavity or nasal passage.[5]

I describe a source/filter model as a metaphor for musical creativity where, analogous to white noise, the source is an acoustic totality. Touching upon the history of 20th-century recorded media and digital information, as well as 21st-century music information retrieval, I hope to trace a narrative of subtractive synthesis as artistic process, complemented by musical examples pointing in possible future directions.

Remix

Since subtractive synthesis emerged in the last century, electronic culture has given increasing emphasis to subtraction as a creative act. Some of the most over-used terms in recent media vocabulary suggest this shift, such as "curator," "DJ," "remix," and "mash-up." Consumption replaces creation among a recent generation of "prosumers."[6] Beyond curated clothing or bookstores, the artistic curator has also grown increasingly visible,[7] as in Carolyn Christov-Barkargiev's curator-driven dOCUMENTA(13).[8] The term "disc jockey," from its first printed appearance in 1941,[9] moved from radio announcers, through hip-hop, then rave, to a broad designation of large numbers of current electronic musicians. Each change has increased the term's creative cachet.

---

[5] Slawson, *Sound Color*, 26. See also Aaron Cassidy's chapter "Noise and the Voice" in this volume.

[6] Dayna Tortorici, "You Know it When You See It," in *What Was the Hipster?*, ed. Mark Greif, Kathleen Ross, and Dayna Tortorici (New York: n+1 Foundation, 2010), 123.

[7] Tom Morton, "A brief history of the word 'curator,'" *Phaidon*, September 9, 2011, http://de.phaidon.com/agenda/art/articles/2011/september/09/a-brief-history-of-the-word-curator/.

[8] dOCUMENTA(13) exhibition website, accessed May 6, 2013, http://d13.documenta.de/.

[9] Marc Fisher, *Something in the Air: Radio, Rock, and the Revolution that Shaped a Generation* (New York: Random House, 2007), 13.

SUBTRACTIVE SYNTHESIS: NOISE AND DIGITAL (UN)CREATIVITY

A predecessor to the DJ was the record collector who, as Jacques Attali demonstrates, can stockpile sound beyond any proportion to listening "use-time."[10] The computer and internet have transformed the DJ and collector further, as software and hardware from iTunes to iPod to Spotify have increased their reach by many orders of magnitude. Yet when creator becomes consumer, so too can listener become "composer," as Attali predicts we will "create our own relation with the world."[11] This is Jacques Rancière's emancipated spectator:

> being at once a performer deploying her skills and a spectator observing what these skills might produce in a new context among other spectators. ... It requires spectators who play the role of active interpreters, who develop their own translation in order to appropriate the 'story' and make it their own story.[12]

Saturation

While the term remix originated from popular music, the visual arts and writing have also embraced it. Christian Marclay draws on 100 years of cinema history in his 24-hour-long *The Clock*.[13] Ai Weiwei combines 1-minute videos taken from 88 highway bridges in *Beijing: The Second Ring & The Third Ring*.[14] Kenneth Goldsmith draws moving portraits entirely from

---

10 Jacques Attali, *Noise: The Political Economy of Music*, trans. Brian Massumi (Minneapolis: University of Minnesota Press, 1985), 101.

11 Attali, *Noise*, 134.

12 Jacques Rancière, *The Emancipated Spectator*, trans. Gregory Elliott (London: Verso, 2009), 22.

13 Christian Marclay, White Cube Gallery website, accessed May 14, 2013, http://whitecube.com/artists/christian_marclay/.

14 Adrian Blackwell, "Ai Weiwei: Fragments, Voids, Sections, Rings," *Archinect*, December 5, 2006, accessed July 30, 2013, http://archinect.com/features/article/47035/ai-weiwei-fragments-voids-sections-and-rings.

transcribed radio archives in *Seven American Deaths and Disasters*.[15] Jonathan Lethem casts a "copy-left" argument almost entirely using quotations from other writers in "The Ecstasy of Influence: A Plagiarism."[16] Or Ai Weiwei inverts industrial re-production by commissioning 100 million hand-made porcelain copies of sunflower seeds.

Figure 1: Ai WeiWei, *Sunflower Seeds*, Tate Modern, 2010 (photo by Mike Peel)[17]

These works have in common not only their borrowings but also their presentation of an archive-like totality, inviting the audience to make their

---

15 Kenneth Goldsmith, *Seven American Deaths and Disasters* (Brooklyn: powerHouse Books, 2013).

16 Lethem, "The Ecstasy of Influence," 93–120.

17 "Ai Weiwei's Sunflower Seeds, Tate Modern" Wikimedia Commons, accessed May 8, 2013, http://commons.wikimedia.org/wiki/File:Ai_Weiwei%27s_Sunflower_Seeds,_Tate_Modern_1.jpg.

SUBTRACTIVE SYNTHESIS: NOISE AND DIGITAL (UN)CREATIVITY

own selections. Attali writes in *Noise*, "the absence of meaning ... is nonsense; but it is also the possibility of any and all meanings,"[18] reminiscent of the random writing machines Jonathan Swift depicts in *Gulliver's Travels*, or Jorge Luis Borges's "The Library of Babel." This idea of noise as total possibility is also cited by recent sonic artists: JLIAT enumerates the total number of possible audio CDs, and Johannes Kreidler says "every piano tone can be understood as a fragment of piano music," as filtered from the historical record.[19] But the idea also goes far back into pre-digital music.

Cage quotes Debussy, "I take all the tones there are, leave out the ones I don't want, and use all the others," and Peter Ablinger relates this to diverse strains of noise in the 20th century.[20] When Cage filters with a radio frequency dial in *Imaginary Landscape No. 4*, or a phonograph cartridge in *Cartridge Music*, the source is the media noise radiating around us, and the composer/performer is an early DJ.[21]

Stockhausen's *Mikrophonie I* is another kind of live subtractive synthesis, with the tam-tam as a noise source filtered, amplified, and focused through attentive listening by composer, performer, and audience. Stockhausen phrases his attitude more generally:

> Use all the components of any given number of elements, don't leave out individual elements, use them all with equal importance and try to find an equidistant scale so that certain steps are no larger than others. It's a spiritual and democratic attitude toward the world.[22]

---

18 Attali, *Noise*, 122.

19 See James Whitehead's chapter "Un-sounding Music" in this volume and Johannes Kreidler, "Music with Music," lecture at the Darmstadter Internationale Ferienkurse, 2010, accessed July 4, 2013, http://www.youtube.com/watch?v=BsFtHsxvoWs.

20 See Peter Ablinger's chapter "Black Square and Bottle Rack" in this volume. Fittingly, Ablinger uses filtered white noise in his own work, for example *Weiss/weisslich 27 "komplementäres Rauschen"* and *IEAOV*.

21 Paul Hegarty, *Noise/Music: A History* (New York: Continuum, 2007), 26.

22 Jonathan Cott, *Stockhausen: Conversations with the Composer* (New York: Simon and Schuster, 1973), 101.

NOISE IN AND AS MUSIC

Other serial or algorithmic techniques, like those used by Brian Ferneyhough, generate a saturation of available materials. The composer is engaged constantly as a listener evaluating, discarding, or incorporating the results, like a photographer examining images in a contact sheet to select the final print. Or in the "saturation" music of composers like Franck Bedrossian, the listener is engaged as a creator, making his or her own selection from the tangled surface.

Figure 2: Toni Frissell, contact sheet, "Fashion model posing near Tidal Basin with Washington Monument in background," 1946[23]

Heard in this way, the expansion of Morton Feldman's late work, or the work of many sound installation artists, asks the listener to trace a path in time and space, filtered through his or her body and attention to shape the

---

23 Toni Frissell (public domain), image from Wikimedia Commons, accessed May 6, 2013, http://commons.wikimedia.org/wiki/File:Toni_Frissell,_fashion_model_near_tidal_basin,_Washington,_D.C.,_ca._1946.jpg.

experience. Like an abstract-expressionist canvas viewed close-up (Rothko specifies a distance of eighteen inches[24]), the viewer's field of vision becomes a filter from which the expanse of the canvas is subtracted.

Richard Barrett's phrase "radically idiomatic," adapted from Derek Bailey's "non-idiomatic" improvisation,[25] could be used to describe many musicians working with noise-rich playing techniques. Like Helmut Lachenmann's *musique concrète instrumentale*, which references Pierre Schaeffer's phenomenological taxonomy, they work with the totality of sound production methods available through the performer's body and instrument, both idiomatic and not. Again composition, performance, and listening are blurred,[26] as playing techniques must be "performed"—through self-experimentation or collaboration—before they are "composed." Selection is here an empirical process, like Schaeffer's "new conditions of observation."[27]

Searchability

This empiricism, however, is experiencing a radical change in the early 21st century as saturation gives way to searchability. In *Uncreative Writing*, Kenneth Goldsmith illustrates how the collage, archive, and cut-up of the 20th century were fundamentally transformed by digital text.[28] Copying is native to the computer, where it is a nearly instantaneous activity as

---

24 Marjorie B. Cohn, *Mark Rothko's Harvard Murals* (Cambridge: Harvard University Art Museums Center for Conservation and Technical Studies, 1988), 15.

25 Richard Barrett interviewed by Daryl Buckley, 2003, accessed May 6, 2013, http://richardbarrettmusic.com/DARKMATTERinterview.html.

26 Or described more generally by the term "musicking" in Christopher Small, *Musicking: The Meanings of Performance and Listening* (Middleton: Wesleyan University Press, 1998).

27 Pierre Schaeffer, "Acousmatics" in *Audio Culture*, ed. Christoph Cox and Daniel Warner (New York: Continuum, 2004), 76–81.

28 Kenneth Goldsmith, *Uncreative Writing* (New York: Columbia University Press, 2011).

documents are automatically cached, downloaded, or cc'd.[29] Text searching has existed as long as digital text itself, or longer, and is one of the primary modes of digital reading.[30]

According to Goldsmith, "with the rise of the Web, writing has met its photography."[31] Perhaps writing at the turn of the last century even borrowed the metaphor of the copy from sound and image media, which had been transformed by recording and photography, respectively. Industrial technology not only enabled the widespread distribution of audio recording but also electrical amplification and broadcast of sound of unprecedented amplitude and reach, as Luigi Russolo celebrates in *The Art of Noise*, now in its centennial year.

Debussy, who himself acknowledged that "the century of airplanes has a right to its own music,"[32] asks Russolo: "can it ever really compete with that wonderful sound of a steel mill in full swing."[33] Russolo, however, stresses that *"the art of noises must not be limited to a mere imitative reproduction."*[34] Further:

> The variety of noises is infinite. We certainly possess nowadays over a thousand different machines, among whose thousand different noises we can distinguish. With the endless multiplication of machinery, *one day*

---

29 Kenneth Goldsmith, *On Uncreative Writing*, lecture at Transmediale BWPWAP, Berlin, February 2, 2013, http://www.transmediale.de/content/kenneth-goldsmith.

30 James Gleick, *The information: a history, a theory, a flood* (New York: Random House, 2011).

31 Goldsmith, *On Uncreative Writing*.

32 Howard Decker McKinney and William Robert Anderson, *Music in History: The Evolution of an Art* (New York: American Book Co., 1957), 640.

33 Claude Debussy, *Writings on Music* (London: Secker & Warburg, 1977), cited in Hegarty, *Noise/Music*, 14.

34 Luigi Russolo, *The Art of Noise (futurist manifesto, 1913)*, trans. Robert Filliou (New York: Something Else Press, 1967), reprinted by ubuclassics, accessed May 4, 2013, http://www.ubu.com/historical/russolo/index.html, 9 (original emphasis).

we will be able to distinguish among ten, twenty or thirty thousand different noises. We will not have to imitate these noises but rather to combine them according to our artistic fantasy.[35]

For Russolo, imitation of noises is only an initial step. What is lacking is the ability to "distinguish" among them: to measure, organize, retrieve. In 1913 the tools for audio searchability were still awaited.

Unlike writing, the transition of sound from analog to digital was not accompanied by a similar revolution in searchability. Archives of digital audio may include many more recordings in much less physical space, but the means of retrieval at first did not fundamentally change. Audio can be searched through textual metadata, an especially powerful tool when it is used in conjunction with the internet. But this is an adaptation from the media of reading and writing, much like the textual copy was imported from the medium of audio recording a century before. Only more recently has technology emerged for searching sound itself, and it could yet transform how we imagine listening and composing.

Concatenation

Since the late 1990s, the growing field of Music Information Retrieval (MIR) has explored the uses of audio feature extraction to summarize information about digital sound.[36] An audio feature, or descriptor, is any characteristic attributed to audio, such as pitch, loudness, brilliance, or higher-level metadata. MIR is changing the way music is categorized, marketed, and recommended through companies like Pandora and The Echo Nest. It also has applications for creation.

---

35 Russolo, *The Art of Noise*, 12 (original emphasis).

36 Michael Fingerhut, "Music Information Retrieval, or how to search for (and maybe find) music and do away with incipits," *IAML-IASA Congress*, Oslo, 2004, accessed May 6, 2013, http://articles.ircam.fr/textes/Fingerhut04b/. See also Nick Collins's chapter "Noise Music Information Retrieval" in this volume.

Emerging from research in speech synthesis, concatenative synthesis uses audio descriptors to analyze short samples of recorded sound and concatenate them into a synthesized output.[37] An increasing number of tools are available for various computer platforms, including CataRT, Soundspotter, MeapSoft, TimbreID, SCMIR, and AudioGuide. CataRT, developed by Diemo Schwarz, implements corpus-based concatenative synthesis (CBCS), where a database containing a large number of samples, the "corpus," is the source. The novelty of CBCS is that it works through concatenation of audio already recorded in its full detail. Like the subtractive synthesis or granular synthesis of a generation before, a desired result is subtracted or selected from a spectrally rich source. But with searchability now made possible through the comparison of audio descriptors, the selection process itself presents new means of organization. And like subtractive and granular synthesis, CBCS is especially effective at dealing with noisy sources, both of acoustic and electronic origins.

Figure 3 (see Appendix, p.229), a screenshot from CataRT, shows a corpus of vocal samples representing 42 minutes of soprano Amanda DeBoer Bartlett singing through vocal "preparations" including police whistle, kazoo, coffee mug, cardboard tube, and spring drum. The recordings are segmented into short "units" and plotted in two dimensions according to descriptor values averaged over each unit: spectral centroid (the perceptual correlate of which is brightness) along the horizontal axis and spectral flatness (a measure of "tonality," here understood as the opposite of noisiness) on the vertical axis. Each vocal preparation is indicated by color and label printed at the center of the corresponding group of units. It can easily be seen how the preparations filter the vocal noises in different ways, adding an extra layer to the source/filter model of acoustic voice production. For example, the samples performed through spring drum or teapot are positioned at the upper left,

---

37 Diemo Schwarz, "Corpus-based concatenative synthesis," *IEEE Signal Processing Magazine* 24, no.2 (March 2007): 92–104.

indicating a relatively low brightness and high tonality. The police whistle, ceramic whistle, and kazoo fall further toward the lower right, indicating higher brightness and lower tonality (or higher noisiness). The unprepared voice, labeled "amanda," lies near the center of the distribution.

The model of a space of sounds identified with audio features as a metric preceded the technology of digital descriptors. Pitch and harmonic lattices, or *Tonnetze*, were used as models during the 18th and 19th centuries.[38] More recently perceptual models of timbre have been proposed by John Grey[39] and David Wessel[40] based on the listening judgments of experimental subjects. These "timbre spaces" were an inspiration for the development of concatenative synthesis tools and are also a compelling model for their application.

Subtractive Composition

I used the corpus of samples shown in Figure 3 (see Appendix, p.229) to compose both the vocal and electronic parts of *Without Words* for soprano, eleven instruments, and live electronics.[41] The vocal part was transcribed from the electroacoustic samples, producing notated material to be re-interpreted by the performer.[42] Taking the empirical approach to an extreme, all of the sounds that were used in the score are already represented in the corpus, and composition is a process of subtraction from this totality. However, unlike

---

38 Fred Lerdahl, *Tonal Pitch Space* (Oxford: Oxford University Press, 2001), 42.

39 John M. Grey, "Multidimensional perceptual scaling of musical timbres," *Journal of the Acoustical Society of America* 61 (1977): 1270–77.

40 David Wessel, "Timbre space as a musical control structure," *Computer Music Journal* 3, no.2 (1979): 45–52.

41 Written for Ensemble Dal Niente and DeBoer Bartlett in 2012.

42 Aaron Einbond, Diemo Schwarz, and Jean Bresson, "Corpus-based transcription as an approach to the compositional control of timbre," *Proceedings of the International Computer Music Conference*, Montréal (2009): 223–26.

some historical sources used by artists discussed above, here the corpus is also a product of the composer and performer that is subject to their preferences, biases, and abilities. Building the corpus, and then subtracting from it, are both layers of the creative process that contribute to the result.

This is not only "radically idiomatic," but also radically personalized, as with composers like Barrett who work with single performers over long periods of collaboration. The samples record DeBoer Bartlett at a specific time and place, and future re-interpretations recall this setting as the live, amplified soprano blends with electronic re-syntheses of her samples. If the work is performed by a different soprano, it becomes a "cover," a change in timbral profile much like the change in sonic identity implied by a pop music cover. This personalization also provides an advantage for performance practice: having recorded all the samples herself, it is easier for the performer to recall how to produce them again. For guidance, in addition to the performance notes in the score, the performer is provided with recorded examples of her own samples as a mnemonic for re-interpretation.

In *Without Words*, subtraction is applied not only to instrumental and vocal sound but also to a database of textual fragments from other collage-inspired writers including Marianne Moore, Douglas Huebler, Walter Benjamin, Ai Weiwei, and Kenneth Goldsmith. DeBoer Bartlett took the texts as a resource for speaking, whispering, singing, and improvising, both unimpeded and through the vocal preparations listed above. When taken as a corpus for concatenative synthesis, the resulting micro-montage bears an unpredictable relationship to the original texts.

While the origins of most of the phonemes are at first unrecognizable, further repetitions and developments open windows on specific syllables, words, and eventually full lines. For example, in Figure 4, "more in" turns out to be taken from Douglas Huebler's conceptual art mantra: "the world is full of objects more or less interesting: I do not wish to add any more."[43]

---

43 Douglas Huebler, "Untitled Statements (September and December 1968)," in Germano Celant, *Arte Povera* (New York: Praeger, 1969), 43.

SUBTRACTIVE SYNTHESIS: NOISE AND DIGITAL (UN)CREATIVITY

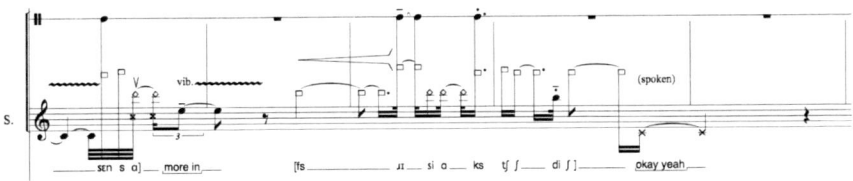

Figure 4: excerpt from *Without Words*, soprano part, mm. 65–69

The other intelligible words, "okay yeah," were ad-libbed by DeBoer Bartlett during the recording session—a trace of the process of collaboration between composer and performer. This "noise" of the sampling procedure, usually edited out, is instead combined in the database on an equal footing with the more literary texts, calling to mind the information theory concept of noise as unwanted signal.

Another source is encoded in the excerpt beyond the texts and vocal samples themselves: the vocal line is an audio mosaic of a field recording. Another metaphor of subtraction—echoing Cage, Ablinger, or R. Murray Schafer and other artists associated with the acoustic ecology movement—the field recording captures noises filtered by the ears and microphones and framed by the start and stop of the recorder. In one of the emerging applications of MIR, an audio mosaic takes a target recording and matches it with a montage of shorter segments from a different audio source. In this case, the spectral descriptors of the vocal samples were matched to the same descriptors of the field recording to assemble a mosaic with the maximum timbral fusion of the two sources. This allows the amplified soprano to fade in seamlessly over the field recording as the work opens. As with the photographer's contact sheet, the choice of recording, corpus, and the specific descriptor weights for the mapping all become compositional decisions.

A field recording of frogs was chosen here, referencing one of the texts in the soprano's database by Marianne Moore: "imaginary gardens with real toads in them."[44] The mapping explains the high brightness values called for in the soprano's whispered speech (indicated by rectangular noteheads), the preparation with police whistle (indicated on the upper staff), and the breathy high Ds (indicated by diamond noteheads), which are a prominent frequency of the frogs' song. The audio mosaicking process is reenacted in performance, as the amplified soprano is accompanied again by her own samples matched according to spectral descriptors in real-time.[45]

Each of the sources activates a different time and place in the work's genesis: the lines written and revised by Moore and Huebler, the "performance" of the frogs captured by the field recording, the soprano's improvised performance during the sampling session, and the live performance in which the soprano reinterprets the transcribed score. Like the memory-infused landscapes in W.G. Sebald's *Die Ringe des Saturn*, multiple temporalities are collapsed as these layers are simultaneously transmitted to the listener.

Arcade/Archive

While historical resonances have long been sourced for creative filtration, whether explicitly musical or not, they also present new possibilities for concatenative synthesis. In *Passagework* for two pianists, two percussionists, and live electronics, written for Yarn/Wire in 2009, the noise of decay is foregrounded with the covered arcades of Paris as a source. Following Walter Benjamin's *Das Passagen-Werk*, field recordings of five arcades in contrasting locations respond to the architecture, time of day, and changing cast of fellow

---

44 Marianne Moore, "Poetry," *Complete Poems* (New York: Penguin Books, 1981), 267. In another layer of transcription, Moore removed this line from the 1924 version in her 1967 *Complete Poems*, in which she notes "omissions are not accidents." See Robert Pinsky, "Marianne Moore's 'Poetry,'" *Slate*, June 30, 2009, accessed May 16, 2013, http://www.slate.com/articles/arts/poem/2009/06/marianne_moores_poetry.html.

45 Aaron Einbond, Christopher Trapani, and Diemo Schwarz, "Precise Pitch Control in Real-Time Corpus-Based Concatenative Synthesis," *Proceedings of the International Computer Music Conference*, Ljubljana (2012): 584–88.

SUBTRACTIVE SYNTHESIS: NOISE AND DIGITAL (UN)CREATIVITY

"improvisers" as well as over a century of time that has left some arcades filled with fashionable shops while others are repurposed or nearly abandoned. Benjamin's *Arcades Project* remains unfinished, as if the strain of pre-digital recopying and rearranging was too much to handle in a work that can be read as a prototype of the hyperlink.

Figure 5: Galerie Vivienne, photographed in 1905 (public domain)

Transcriptions for Yarn/Wire, composed through a process of sampling and mosaicking similar to *Without Words*, reveal these changing urban environments as the musicians' own samples define the spaces inside two pianos. The percussionists, using only mallets and small instruments placed inside the pianos, activate these spaces while the field recordings are played through speakers positioned under the soundboards. For example, in Figure 6, one percussionist uses a bird call to mimic distant birds in a neighboring park, while the other uses homemade foam mallets and fingertips to excite resonances corresponding to an automobile engine outside the arcade's exit.

The analogy between the architectural space of the arcades and the acoustic space of the instruments suggests the work as a performed installation, a sonic sculpture that draws the listener into the delicate soundscape. He or she may strain to identify the sources of the performers' nearly invisible, amplified gestures, as well as the field recordings filtered acoustically by the resonance of the pianos above the hidden loudspeakers, another version of live subtractive synthesis. This threshold between Schaefferian "reduced listening" and causality reveals the antecedents of the work in Parisian *musique concrète* and Berliner *Hörspiel*. Or alternatively the pianos are an archive out of which these historical references, along with the storied instrumental sounds, are drawn by the performers, like the antique wooden card catalogue in Janet Cardiff's and George Bures Miller's *Cabinet of Curiousness*.[46]

Noise Movement

The internet offers a yet noisier source of information for purposive filtration. In *Resistance* for prepared bass clarinet and live electronics, written for Heather Roche, recordings are taken from YouTube of the "Occupy Wall Street" and "Arab Spring" protests of 2011. While referencing possible political meanings

---

46 Janet Cardiff and George Bures Miller, *Cabinet of Curiousness*, 2010, artists' website accessed May 14, 2013, http://www.cardiffmiller.com/artworks/inst/cabinet_of_curiousness.html.

SUBTRACTIVE SYNTHESIS: NOISE AND DIGITAL (UN)CREATIVITY

Figure 6: excerpt from *Passagework*, mm. 199–207, corresponding to field recording of Galerie Vivienne

of noise, as developed by Attali and others, these samples also present unique sonic qualities. The low-fi, compressed, and over-saturated samples become advantageous materials for transcription and cross-synthesis with amplified bass clarinet. The bass clarinetist's sound is also modulated with overblowing and screaming (indicated by encircled noteheads and diagonal lines in the example in Figure 7), distorting it into a noise-rich source. This is combined with protest samples played from a low-quality miniature loudspeaker placed in the bell of the instrument, another literal application of source/filter subtraction through the colored speaker response and instrumental cavity. Preparations with paper, plastic, and aluminum foil over speaker and bell produce an additional layer of filtration and distortion before being further amplified (indicated by cross-hatched horizontal lines in Figure 7). A signal "path" is traced from the performances of the original protesters, through recording and post-processing, through the audio ripped from YouTube, through the miniature loudspeaker in the bass clarinet, blended with the live sound of the instrumentalist and her voice, through a layer of foil, through an additional microphone for amplification, and diffused out of full-range speakers to the listener.

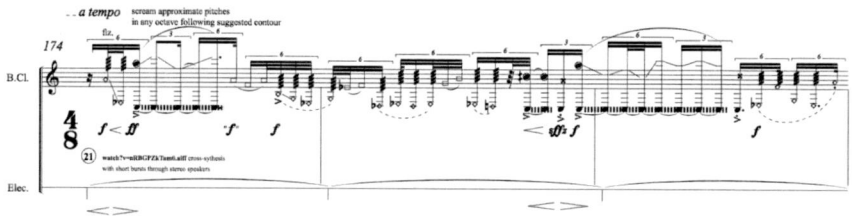

Figure 7: excerpt from *Resistance*, mm. 174–76

Similar to *Without Words*, the excerpt shown in Figure 7 is a transcription of audio from a YouTube video of Tahrir Square, Cairo.[47] This source has been mapped to an audio mosaic of samples of Roche improvising on prepared bass clarinet, then transcribed into notation. The re-interpreted result is cross-synthesized through spectral convolution with the original sample, causing amplification of the spectrum where the two signals overlap, and a filtration where they do not. This produces a controlled feedback loop between the microphone used for processing and the miniature speaker. In this way the residue of imperfections in the transcription process becomes a feature of the saturated texture, forcing another layer of subtraction where the two do not coincide.

These examples suggest just a few possible directions the audio database (personal archive) and internet (collective archive) can take as sources of compositional synthesis. Others are being explored by a young generation

---

47 Al Jazeera English, "Scenes from Tahrir Square: Mubarak's Non-Resignation," accessed May 9, 2013, https://www.youtube.com/watch?v=nRBGPZkTam0.

of composers and improvisers including Johannes Kreidler, Maximilian Marcoll, Matthew Shlomowitz, Alec Hall, Bryan Jacobs, William Brent, Ben Hackbarth, and Diemo Schwarz. Many of these artists program their own computer tools: software development becomes an extension of the compositional process for gathering and filtering material. Kreidler, discussing his work *product placements*, in which 70,200 sources are sampled in a 33-second electroacoustic piece, summarizes: "the work is a network."[48] But even with music bathed in samples, databases, references, and borrowings, the contrasting musical results these artists derive attest to the individuality of their approaches.

Less is more: the future promises to be subtractive.

---

48 In German: "Das Werk ist ein Netzwerk." Thomas Groetz, "'Kunst muss verdächtig sein': Der Komponist und Aktionskünstler Johannes Kreidler," *MusikTexte* 123 (2009): 5–10.

NOISE IN AND AS MUSIC

Pierre Alexandre Tremblay

What is noise (music) to you?

If we take the musical act as expression, therefore communication, then signal theory forces us to accept every sound as music as long as it is intended to be. What is not intended is noise, even in the form of a Bach partita played next door when I want to listen to Bernhard Günter. Music as art is a statement of intent, and everything (and only what) the artist wants to produce is the work. There is only noise in non-intent.

That definition of music versus noise, in line with Varèse's "organised sound," has a contradicting sibling in my practice: I like to embrace uncertainty, the unexpected, and instability fascinates me as it reminds me of living things. Only human thoughts and models can dream up abstract perfection, and I prefer the filthy, gritty, humid, sticky unevenness of nature. So music, as human construct, contrasts with noise as the undesired artefact of reality. In that respect, we could take the harmonicity of a signal to declare it pure, and inharmonicity to be noisy ... quite a boring scale, but quite useful, and definitely in phase with the ideals of consistency in instrument making and performance practice of the Western world. This scale has its limits though, as there are perceptual limits to saturation. One is *Gestalt* grouping, when accumulation becomes too dense; another is that we perceive in contrasts and get used to anything that is constant, even information overload. Therefore I like to think of this axis as being curved, with the maximum noisiness at the peak of perceptual saturation. In this respect, the context of listening becomes essential in the definition of noise itself.

I think we are in a post-noise period. Post-noise, as in post-glitch. In the latter, what was a mistake of digital sound devices was embraced as art, first with its conceptual random value, then the sound of glitch became an object itself. Comparably, we have now heard every sound, loud and soft, saturated and pure, dense and light, with classic noise artists mostly embracing the former in each of these dualities. This soundworld has now permeated all music practices. Our soundscapes are lower-fi than ever, the sound design in multimedia is getting more daring by the day, chart-topping pop music distorts like never before, and the loudness war has reached its theoretical limits. Moreover, as individuals we are saturated with information as never before and with sheer violence at levels that are unattainable in art—radical art is *passé*: it looks quite futile compared to the daily news. More interestingly, the grey zone between these extremes is now fully assumed, full of rich crossovers and hybrids: in-between-ness reigns. This post-noise era is rich with the full breadth of the sonic world.

Maybe the desire of artists to use the dirty part of their soundworld has been there forever; they were just surrounded by cleaner sounds. With a louder world, with omnipresent background music, with piercing sirens on the street, the threshold of what is acceptable as music might have just been pushed. Here is an interesting hint that we might have reached a limit: after the re-appropriation of the soundspace by the individual, with the headphones/Walkman revolution of the 1980s, the trend of active silencing headphones points to something quite clear: noise is in the ear of

Why do you make it?

Am I making noise? Not to my ears, but I certainly embrace all levels of volume, contrast, purity, transience, density, and instability in the listening experience. This might be noise to some, but to me it is music.

I like art to be in phase with the world, and to embrace the rich multitude of different human experiences from an embodied perspective, even if this can be a little overwhelming in its contrasts. To embrace my humanity, be it simple pleasure or existential turmoil, music is the only way to express it: art is for my gut and my soul.

Noise Music Information Retrieval

Nick Collins

Noise music foregrounds processes that other music might seek to minimize or avoid, from high levels of distortion to problematic and polemical subject matter.[1] The focus here is on the typical "noisy" manifestation of extreme sensory dissonance in the audio signal itself, and tracking this through automatic analysis techniques from the field of Music Information Retrieval (MIR).[2] Such MIR methods have strong applications in computational musicology when working with one or more audio files. For example, Collins[3] applies MIR tools to a study of a corpus of synth pop; Klein et al.[4] consider using MIR methods for the analysis of acousmatic electroacoustic music; Tsatsishvili[5] attempts to differentiate the overlapping sub-genres of heavy metal; and Mital and Grierson[6] explore an archive of works of Daphne Oram through a visualization method.

---

1 Paul Hegarty, *Noise/Music: A History* (New York: Continuum, 2007); Thomas Bey William Bailey, *Microbionic: Radical Electronic Music and Sound Art in the 21ˢᵗ Century* (London: Creation Books, 2009); Nick Collins, Margaret Schedel, and Scott Wilson, *Electronic Music* (Cambridge: Cambridge University Press, 2013).

2 Michael A. Casey, Remco Veltkamp, Masataka Goto, Marc Leman, Christophe Rhodes, and Malcolm Slaney, "Content-based music information retrieval: Current directions and future challenges," *Proceedings of the IEEE* 96 no.4 (2008): 668–96.

3 Nick Collins, "Computational Analysis of Musical Influence: A Musicological Case Study Using MIR Tools," *Proceedings of the International Society for Music Information Retrieval Conference, Utrecht* (2010): 177–182.

4 Volkmar Klien, Thomas Grill, and Arthur Flexer, "On Automated Annotation of Acousmatic Music," *Journal of New Music Research* 41 no.2 (2012): 153–73.

5 Valeri Tsatsishvili, "Automatic Subgenre classification of heavy metal music" (Master's Thesis, University Of Jyväskylä, 2011).

6 Parag K. Mital and Michael Grierson, "Mining Unlabeled Electronic Music Databases through 3D Interactive Visualization of Latent Component Relationships," paper presented at New Interfaces for Musical Expression (NIME), Daejeon, South Korea, 2013.

In this project, MIR-informed analysis is explicitly applied to noise music to assess the structure of individual recordings and to compare multiple recordings, providing new insights into the sonic content of noise. The specific targets of this study include two Merzbow albums, *Oersted* (1996) and *Space Metalizer* (1997), for which individual tracks are analyzed, and the pieces across the two albums compared. I also investigate the application of MIR analysis to a corpus of historic noise music, including Whitehouse, Masonna, and Xenakis, placing the Merzbow works in a wider context. Although there is a broad potential to this technology, which may extend beyond musicology to new compositional directions, there are also challenges. Questions remain of how to validate the results of automated analysis against human reaction, and a critical view of MIR should be maintained as we proceed. Nonetheless, the study is an essential step to approaching and evaluating MIR applications.

Noise MIR analysis techniques

MIR typically operates with respect to a corpus of audio files, though it may also examine properties of a single file in isolation. Rather than working with raw sample data, a system will extract derived features, such as energy in chroma (typically following pitch classes representing the semitonal twelve-note equal temperament typical of Western music) or Mel-Frequency Cepstral Coefficients (a measure of spectral energy distribution with good correspondence to timbre). As well as having a lower sampling rate than audio samples, making the amount of data more manageable, these features are chosen because they ideally represent more salient auditory and musical attributes of the work under study. Often in MIR, a time-varying feature is reduced further, perhaps to an average across an entire piece; this can still help to plot locations of different pieces with respect to another. Nevertheless, retaining the dynamical progress (the time series) of features within given pieces can capture musical behavior in more detail and is intuitively closer to a human-like perception over time. Having obtained features—essentially, numerical summaries—for some set of files, machine learning algorithms can then be applied over the corpus to examine such questions as how all the files (music) cluster together, how to discriminate different "types" of file (music), and so forth.

This project favors such features as spectral entropy, sensory dissonance, perceptual loudness, "transientness," and spectral centroid as timbral aspects of high relevance to the perception of noise music. Some of these features, such as sensory dissonance and perceptual loudness, depend on models of human auditory perception; others are related to studies of instrumental timbre (correlating for the spectral centroid to "brightness" of tone), or are information theoretic signal processing constructions (spectral entropy is a measure of the information gained in momentary spectral change). I will use both summary features such as a mean (an average) across a whole piece, and time-varying features that describe the course of a piece, as explained further below.

Once features have been extracted, similarity measurements can compare within-piece or between-piece relations, examining the internal structure of a work, or the proximity of different audio files representing different opuses from the same or diverse composers. Similarity matrices can help assess within-piece formal relationships, including the detection of change points through a derived novelty curve. Models can be formed for individual pieces with respect to their time series, for instance, via k-means clustering from the feature vector space to cluster labels, followed by variable order Markov modeling on the symbolic sequences created. Feature statistics, or time series models, can then form the basis of comparison between pieces in corpus analysis.[7] All calculations in this study use the open source SCMIR library for SuperCollider.[8]

---

[7] It is beyond the scope of this chapter to cover MIR techniques in general. For an introduction to content analysis, see Casey et al. "Content-based music information retrieval."

[8] Nick Collins, "SCMIR: A SuperCollider Music Information Retrieval Library", *Proceedings of the International Computer Music Conference, Huddersfield* (2011): 499–502.

As an example of feature extraction, Merzbow's "cover" of *Silent Night* from a noise music Christmas compilation[9] was subjected to analysis. "Cover" is apt, as Merzbow gradually covers a strained rendition of the carol in layers of noise until the source is completely obscured. With this increase in obscuration under a wall of noise, the track provides a useful test for whether computational feature extraction discovers the progression into brutal noise clearly audible by a human listener.

Figure 1 plots how feature values vary as the initial carol submerges beneath the distortion. In part, this plot provides a validation of the types of features being extracted; they all generally show an increase that follows the highly perceptible increase in distortion as the piece progresses, though moment to moment variation is also evident. (The plotted values are maximal values for each second of audio; the FFTCrest feature, which looks at spectral "peakiness," is inverted so that the flatter spectrum of the noisier portions is evident.) There are short silences at the beginning and end of the track that disrupt feature collection a little (the SpectralEntropy in particular is maximal in the face of silence, explaining its later relatively low values, though the upwards trend is still evident).

It is also plausible to detect abrupt transitions in the time varying feature values to seek out potential sectional boundaries within a work. This may be done with respect to a single feature, or over multiple features simultaneously, for example via structural detection methods such as convolution with a checkerboard kernel along the diagonal of a similarity matrix. A 'findSections' command in SCMIR pointed to one main transition point, at 92 seconds in; this turned out to correspond to a sudden increase in the distortion and brightness of the guitar. While the routine did not reveal any additional points of change, the other layers tend to come in more slowly, or to reflect further levels of distortion that the feature detection may not fully track.

---

9 *Various–The Christmas Album* (Sony Records, SRCL 3723, 1996), accessed July 29, 2013, http://www.discogs.com/Various-The-Christmas-Album/release/1331008.

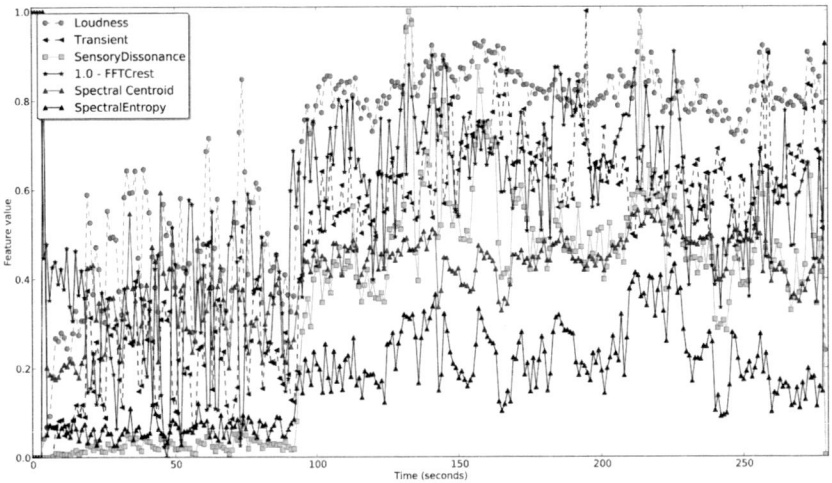

Figure 1: Feature trails over the four and a half minutes of Merzbow's *Silent Night* (1996)

The change point is also evident from the visual tipping point of the feature curves in Figure 1. I do not pursue change points further in this article, but the technique illustrates the further options for structural analysis opened up by MIR.

Merzbow vs. Merzbow

To illustrate application across multiple files, this section explores the similarity relations among the tracks over two Merzbow albums, *Oersted* (1996, 4 tracks titled by their durations) and *Space Metalizer* (1997, five tracks). Products of a period in Merzbow's 1990s output associated with harsh noise, the sound sources include various classic analog synthesizers, such as the EMS Synthi A, and metal percussion, running through effects units such as guitar pedals and filters alongside live tape manipulation. There is a feeling of spontaneous performance across the tracks, with mercurial shifts in timbre as attention wanders across the equipment at the artist's disposal. *Oersted* is perhaps a little starker, yet both albums have free time moments as well as some tight repeating loops evident at points. Both push hard, with hot mastering.

Machine analysis initially extracted six features: perceptual loudness, transientness (a measure of sudden signal change, for example caused by percussive onsets), sensory dissonance, a spectral crest measure (the "peakiness" of the spectrum), spectral centroid, and spectral entropy. All features were minimum–maximum normalized to the range 0.0 to 1.0 with respect to a globally derived maximum and minimum over all files. Means and maximums were taken across complete files as summary features, as well as one-second window means and maximums as time-varying features.

As a sanity check on the feature extraction, it is clear from the feature data that *Oersted* is louder and has higher sensory dissonance (according to the computational models) than *Space Metalizer*, somewhat confirmed by listening (all the tracks are mastered hot and general loudness is persistent, though with moment to moment fluctuation). The greatest maximum sensory dissonance of the tracks was for *7:53* on *Oersted*, the most unrelenting in its wideband noise; the lowest was "Son of Zechen" from *Space Metalizer*, which is slightly gentler (relatively!) on listening. It is clear from examination of the feature trails (time-varying one-second means and maximums) that features are often (though not inevitably) correlated; for example, loudness, transientness, and sensory dissonance all react to the sheer "energy" (in a non-technical sense) of the music.

The tracks were compared in two ways. The first simply looked at the proximity of the six mean feature values, a cruder measure. The second used time series methods, to be described below, to form a model of each piece; once a model was formed, it could be used to predict how surprising other pieces appeared with respect to that prior knowledge, so as to measure similarity.

The first mean-based method created the similarity matrix in Figure 2. Distances between the six-dimensional feature vectors (the means over a given piece for each of six features) were calculated via the Euclidean metric, and were then normalized across all distances to the range 0–1 for ease of reading (numbers are taken to 3 decimal places). This matrix is symmetric, with zeros on the diagonal, as a piece is always no distance from itself. The rows and columns allow a measure of similarity to be read between any two tracks across the two albums. (The two albums are denoted by prefixes OE and SM.)

	OE1	OE2	OE3	OE4	SM1	SM2	SM3	SM4	SM5
OE1	0	0.383	0.245	0.047	0.131	0.212	0.102	0.191	0.101
OE2	0.383	0	0.021	0.284	0.478	1	0.321	0.09	0.247
OE3	0.245	0.021	0	0.185	0.36	0.812	0.216	0.042	0.161
OE4	0.047	0.284	0.185	0	0.045	0.266	0.028	0.103	0.055
SM1	0.131	0.478	0.36	0.045	0	0.266	0.026	0.195	0.096
SM2	0.212	1	0.812	0.266	0.266	0	0.379	0.694	0.432
SM3	0.102	0.321	0.216	0.028	0.026	0.379	0	0.086	0.04
SM4	0.191	0.09	0.042	0.103	0.195	0.694	0.086	0	0.08
SM5	0.101	0.247	0.161	0.055	0.096	0.432	0.04	0.08	0

Figure 2: Similarity matrix over two Merzbow albums from mean feature vectors

There are overlaps between the two albums, given their close gestation in time within Merzbow's career. SM2, "Son of Zechen," is the most atypical track here, seen as particularly different to the central tracks on *Oersted*, though of equal distance to OE4 and SM1. It is closest to the first track of *Oersted*, which may be related to the use of a low, throbbing, bassy figuration in both tracks. The matrix may point to greater variety on *Space Metalizer* than on *Oersted*, in the sense of the degree to which tracks within the album are dissimilar to one another. Yet if *Oersted* is a little more homogenous, this seems predominantly due to SM2 on the other album. With "Son of Zechen" excluded, *Oersted* may be the more heterogeneous; for instance, the bottom-right 3 by 3 submatrix shows the close proximity of SM3 through SM5 within *Space Metalizer*, and the low scores on the SM1 row compliment this picture.

In the second method of assessing similarity, machine analysis proceeded as follows:

1. Form one-second windows of features, taking the mean in each window.
2. Vector quantize; k-means clustering is applied with 20 cluster centers over all mean feature vectors (from all seconds of all pieces) in the 6-dimensional feature space. All (continuous) mean feature vector sequences are mapped to (discrete) integer sequences.
3. Train prediction by partial-match variable-order Markov models for each audio file in the corpus based on the integer sequences.[10]
4. For each pair of files, calculate similarity based on the symmetric cross-likelihood.[11] The model from piece A is used to predict the unexpectedness of piece B, and from B to predict A; the result is a combined sense of how well A and B predict each other and, thus, their similarity.

The result shown in Figure 3 is a similarity matrix between pieces that can potentially illuminate the proximity of the musical thinking of different tracks, as it respects the time variation within those tracks far better than a gross average.

In general outline, the two matrices constructed via two different methods show similar inter-relations between tracks, which may give some confidence to both methods' applicability to noise music. In the second matrix, SM1 and OE3 are the most dissimilar, though SM2 and OE2 (the furthest apart in the

---

[10] Nick Collins, "Influence in Early Electronic Dance Music: An Audio Content Analysis Investigation", *Proceedings of the International Society for Music Information Retrieval Conference, Porto* (2012): 1–6 ; Marcus Pearce and Geraint Wiggins, "Improved methods for statistical modelling of monophonic music," *Journal of New Music Research 33*, no.4 (2004): 367–85.

[11] Tuomas Virtanen and Marko Helén, "Probabilistic model based similarity measures for audio query-by-example," *Proceedings of the IEEE Workshop on Applications of Signal Processing to Audio and Acoustics*, New York (2007): 82–85.

	OE1	OE2	OE3	OE4	SM1	SM2	SM3	SM4	SM5
OE1	0	0.758	0.748	0.703	0.843	0.615	0.724	0.744	0.721
OE2	0.758	0	0.596	0.615	0.939	0.835	0.76	0.671	0.692
OE3	0.748	0.596	0	0.744	1	0.766	0.812	0.682	0.747
OE4	0.703	0.615	0.744	0	0.667	0.589	0.694	0.685	0.659
SM1	0.843	0.939	1	0.667	0	0.696	0.567	0.861	0.612
SM2	0.615	0.835	0.766	0.589	0.696	0	0.539	0.769	0.697
SM3	0.724	0.76	0.812	0.694	0.567	0.539	0	0.718	0.651
SM4	0.744	0.671	0.682	0.685	0.861	0.769	0.718	0	0.584
SM5	0.721	0.692	0.747	0.659	0.612	0.697	0.651	0.584	0

Figure 3: Similarity matrix from predictive models over two Merzbow albums

first matrix) are not conversely claimed as similar. Potentially problematic in the second matrix is the low similarity between the tracks "Space Metalizer Pt. 1" and "Space MetalizerPart 2" (*sic*; this is how the track names are written in the CD liner notes). However, on listening, the difference between Parts 1 and 2 is clear as the tracks proceed; they seem to start at a common point, then diverge into their own noise worlds.

Taking a cue from some of Masami Akita's own preoccupations, one interesting facet of this study in the context of noise music is a measure of "dominance," by which a predictive model trained on one work explains another work better than the other way around. In order to calculate this, instead of the symmetric measure we can use the direct prediction scores (expressed as average logloss, where low numbers denote better predictions). The difference between B predicting A and A predicting B is a measure of the degree to which A dominates B and vice versa. An anti-symmetric matrix is presented here where 1 means that the left row track dominates the column track, -1 is the inverse, and 0 indicates that there is no dominance relation (for the self-model predictions on the diagonal).

	OE1	OE2	OE3	OE4	SM1	SM2	SM3	SM4	SM5
OE1	0	1	1	-1	1	1	1	1	1
OE2	-1	0	-1	-1	1	1	-1	1	1
OE3	-1	1	0	-1	1	1	-1	1	1
OE4	1	1	1	0	1	1	1	1	1
SM1	-1	-1	-1	-1	0	-1	-1	-1	1
SM2	-1	-1	-1	-1	1	0	-1	1	1
SM3	-1	1	1	-1	1	1	0	1	1
SM4	-1	-1	-1	-1	1	-1	-1	0	1
SM5	-1	-1	-1	-1	-1	-1	-1	-1	0

Figure 4: Dominance matrix over two Merzbow albums

According to this measure, track OE4 ("18:49") is the most predictive of other tracks; we might view it as somehow encapsulating a kernel of techniques used throughout the two albums. It is probably the most timbrally varied track on *Oersted*, and it also has an aural relation to many moments on *Space Metalizer*. SM5 ("Mirage") is the most derivative by this measure (all -1s on its row). And Part 2 of "Space Metalizer" dominates Part 1. However, these results should be set in the context of the simplification, whereby actual numerical differences have been reduced to one of three options, and are critically dependent on the aural validity of the modeling in the first place.

A small historical corpus of noise music

I now place these two albums by the same artist in the context of music by other noise musicians. The survey I present is by no means exhaustive; Merzbow's career from 1979 is itself replete with many more recordings than are examined here, and the reader will no doubt think of many other examples of noise musicians that she or he might be interested to explore.

One justification for carrying out this study, even with some reservations on the acuity of features raised in the preceding sections, is that it is impossible for an analyst to keep in mind all of the audio material across a large corpus. Automated methods at least objectify the process of hunting for interrelations on a level playing field, rather than the most recently consulted track, or the bias of a musicologist having listened more to particular material. Nonetheless, the choices made in order to establish a corpus and the decisions taken in writing the computer program are themselves a potential source of bias, if more explicitly stated.

Figure 5 lists the various sources in the corpus gathered for this part of the study. These range from electronic music by Iannis Xenakis, through the classic Lou Reed statement of feedback *Metal Machine Music*, to postpunk experimental acts such as Whitehouse who started to foreground blasts of noise, the Japanese noise artist Masonna, and a control case of the Beach Boys. The total audio content is around 10 hours of material over 96 individual files.

I applied the same processing as above, extracting the same six features with respect to the same normalization factors derived from the Merzbow

Group ID	Artist	Works, with Dates	Total Duration (minutes)
0	Merzbow	*Oersted* (1996), *Space Metalizer* (1997)	135.8
1	Xenakis	*Bohor* (1962), *Taurhiphanie* (1987), *Gendy3* (1991), *S.709* (1994)	58.5
2	Lou Reed	*Metal Machine Music* (1975)	64.2
3	Whitehouse	*Birthdeath Experience* (1980), *Great White Death* (1985)	73.2
4	Nurse with Wound	*Chance Meeting on a Dissecting Table of a Sewing Machine and an Umbrella* (1979)	47.9
5	Einstürzende Neubauten	*Strategies Against Architecture, Vol. 1* (1983)	41.3
6	Non	*Easy Listening for Iron Youth: The Best of Non* (1989)	67.5
7	Masonna	*Shock Rock – Track 2* (2002), *Mademoiselle Anne Sanglante Ou Notre Nymphomanie Auréolé* (1993), *Shinsen Na Clitoris Part 1* (1990)	55.4
8	Beach Boys	*Pet Sounds* (1966)	36.4

Figure 5: Noise music corpus

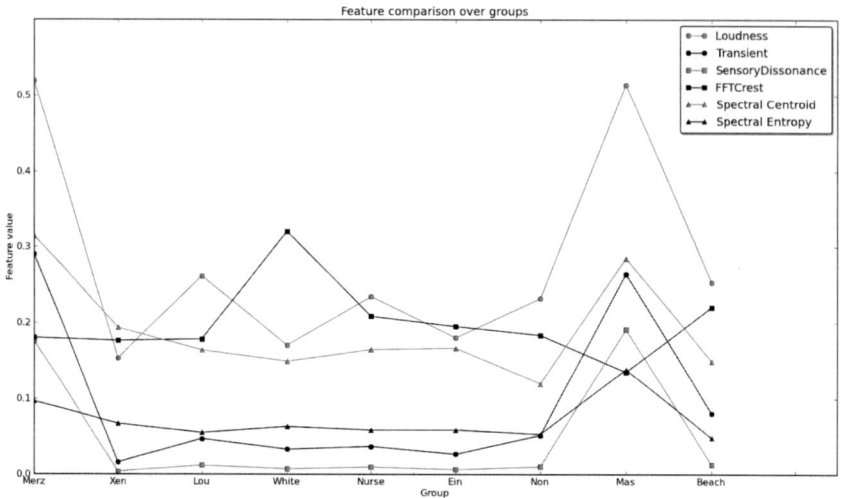

Figure 6: Mean feature values by artist group for the corpus in Figure 5

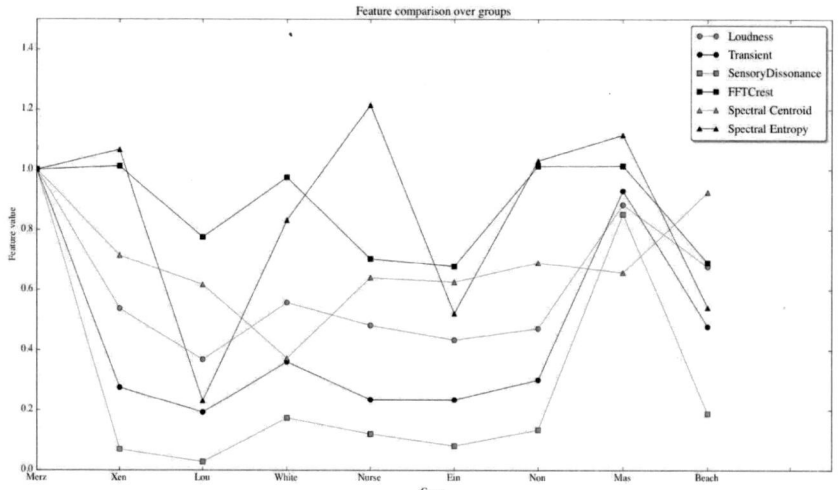

Figure 7: Maximal feature values by artist group for the corpus in Figure 5

albums; feature values are thus relative to the Merzbow group as the reference. All tracks from a particular artist form a group of feature data, giving nine distinct musical groups and their associated data.

Figures 6 and 7 plot the mean and the maximal feature values for each artist group over the six extracted features. Note that group 0, the Merzbow reference, has every maximum at 1, as it provided the normalization information for the feature extraction. These diagrams clearly show the relationship between the Merzbow and Masonna groups of tracks. Perhaps surprising is that Metal Machine Music does not enter so similar a space; in actual fact, Lou Reed's opus is much more based around resonant ringing pitches of feedback, and has a less broadband noise characteristic and less fierce mastering. Because of the percussion transients, the Beach Boys album takes a closer aspect (with respect to the extracted features) to other noise works than might be expected. The two Japanoise artists are clearly revealed as more aggressive in their soundworld than some earlier noise precedents.

The interrelationship of pieces was assessed via the feature-based similarity measurements described above, first using means and maximums over each group, and then with time series models trained over all the pieces within a given group. While this might also be investigated on a piece-by-piece basis, here it is restricted to artist groupings.

Using only group feature means and a Euclidean metric:

	Merz	Xen	Lou	White	Nurse	Ein	Non	Mas	Beach
Merz	0	0.962	0.669	1	0.748	0.895	0.781	**0.021**	0.652
Xen	0.962	0	0.052	0.088	0.034	**0.008**	0.05	0.916	0.07
Lou	0.669	0.052	0	0.11	**0.007**	0.028	0.011	0.631	0.012
White	1	0.088	0.11	0	0.064	**0.061**	0.09	0.999	0.074
Nurse	0.748	0.034	**0.007**	0.064	0	0.012	0.011	0.716	0.01
Ein	0.895	**0.008**	0.028	0.061	0.012	0	0.021	0.855	0.035
Non	0.781	0.05	**0.011**	0.09	**0.011**	0.021	0	0.736	0.013
Mas	**0.021**	0.916	0.631	0.999	0.716	0.855	0.736	0	0.633
Beach	0.652	0.07	0.012	0.074	**0.01**	0.035	0.013	0.633	0

Figure 8: Similarity matrix over corpus from mean feature vectors

Using models trained on the work of a particular artist, and used to predict other artists:

	Merz	Xen	Lou	White	Nurse	Ein	Non	Mas	Beach
Merz	0	0.927	0.86	0.92	0.925	1	0.787	**0.401**	0.838
Xen	0.927	0	0.323	0.294	0.322	**0.266**	0.303	0.706	0.542
Lou	0.86	0.323	0	0.407	0.372	0.384	**0.213**	0.47	0.376
White	0.92	0.294	0.407	0	0.404	0.486	**0.214**	0.614	0.497
Nurse	0.925	0.322	0.372	0.404	0	**0.215**	0.233	0.627	0.408
Ein	1	0.266	0.384	0.486	**0.215**	0	0.292	0.651	0.502
Non	0.787	0.303	**0.213**	0.214	0.233	0.292	0	0.528	0.349
Mas	**0.401**	0.706	0.47	0.614	0.627	0.651	0.528	0	0.733
Beach	0.838	0.542	0.376	0.497	0.408	0.502	**0.349**	0.733	0

Figure 9: Similarity matrix over corpus from predictive models

The closest proximity of artists in each row is indicated in bold type. Merzbow and Masonna are paired on both measures, while they are dissimilar to other artists in the corpus. Interestingly, Lou Reed appears much closer to Masonna than Merzbow, although Masonna is still the second most distant from him. Some overlap among experimental acts of the later 1970s and early 1980s is apparent, for instance, in the close relation of Nurse with Wound and Einstürzende Neubauten.

	Merz	Xen	Lou	White	Nurse	Ein	Non	Mas	Beach
Merz	0	1	-1	1	-1	-1	1	1	1
Xen	-1	0	1	-1	-1	-1	-1	-1	-1
Lou	1	-1	0	-1	-1	-1	-1	-1	-1
White	-1	1	1	0	-1	1	1	-1	1
Nurse	1	1	1	1	0	1	1	-1	-1
Ein	1	1	1	-1	-1	0	1	-1	-1
Non	-1	1	1	-1	-1	-1	0	-1	-1
Mas	-1	1	1	1	1	1	1	0	1
Beach	-1	1	1	-1	1	1	1	-1	0

Figure 10: Dominance matrix over corpus

The dominance matrix technique can also be applied over the artist groups and their models, as shown in Figure 10. This dominance table should not be taken too seriously and is included here as much as a provocation to consider the side effects of accepting technological musicology uncritically. Nonetheless, Merzbow dominates Masonna, and Lou Reed dominates Merzbow (but no other artist; the dominance relation is not transitive, as Lou Reed is subservient to Masonna). The Beach Boys dominate many other artists; this is likely an artifact of their (possibly) broader spectral palette and time variation with respect to the more delimited timbral world of some noise music artists.

In summary, the aggressive sound of Japanese noise music from Merzbow and Masonna is clearly indicated. However, the surprising relation of the Beach Boys to some of the other artists represented here gives pause; the combination of features extracted is probably not sufficiently representative. It certainly doesn't differentiate the common practice harmony and rhythm that appears in the Beach Boys' work.

Conclusions

Some analytical applications of MIR techniques, such as feature extraction and similarity measures, have been investigated with respect to noise music. While such procedures may reveal new formal details and new interrelations of pieces within a systematic framework, there remains a need to validate machine listening against human listening. Such analysis will remain a companion and compliment to the human analyst, but its use should be further investigated as MIR audio analysis techniques continue to develop.

I have said less about compositional applications, which are themselves at the mercy of the quality of algorithmic listening. Ideally, audio analytical methods can form the basis for critical systems for algorithmic composition and interactive music systems. For a noise music generating system, analytical techniques can provide the grounding, the listening experience, for a self-critical computer agent, whether founded on a composer's own work, or a database derived from a historical or contemporary corpus. Given the "otherness" of noise music, an exact simulacrum of human listening may not necessarily be an aesthetic requirement, and partially effective auditory modeling may allow for many noisy possibilities.

The challenges for future work are many. There is the problematic nature of many noise musicians' release catalogues, which embrace prodigious release rates over many media as a sort of cultural noise;[12] serious Merzbow or noise music history scholars may be forced to pool resources and form collaborative online databases, with associated issues of copyright. Although a musicologist's own listening is a natural centerpiece of analysis projects, the computer acts as a proxy listener, necessarily as the corpus size increases. Engaging with the vast extent of noise music releases may require computational assistance, and indeed, certain aspects of noise music structure on the boundary of human discernment may benefit from untiring neutral signal processing analysis. However, the extent to which computer extracted features reflect the encultured and embodied human auditory system remains an open question; I have provided some validation here when the computer has suggested particular conclusions, but much validation remains to be done across many musical arenas. The relationship of the Beach Boys to other works in the corpus study is in part problematic because further features working for general popular music may be required, if only to discount numerically their relevance to much noise music practice.

Noise music can reveal novel issues with feature extraction itself; for example, in this project, where normal sensory dissonance levels in popular music had not previously overloaded the detector, the threshold had to be reset to cope with Merzbow without registering a constant clipped value! The choices of thresholds of detection in the face of a wide variation of signal to noise ratios illustrates again some disconnection from a truly human-like listening experience. There may be earplugs and the stapedius reflex, or a degree of cultural learning about the possibilities of noise music, but the human auditory system does not require literal rewiring to cope with it; instead, noise music can exploit certain timbral limits of a pre-existing auditory biology. Should a computational system learn to appreciate greater

---

12 Matthew Blackwell, "In Left Field: Merzbow's Discography–Noise Music and the Taxonomic Drive" (Dec 7, 2011), accessed July 29, 2013, http://www.prefixmag.com/features/john-wiese-merzbow/merzbows-discography/59054/.

depths in noise music with repeated exposure, in order to model the route of human listeners?

Future computational musicology of noise music might use far larger databases that better reflect the diversity and mass release schedules of noise music across various currents of experimental and counter-culture work (consider the relationships with extreme metal, electronic body music and industrial acts, breakcore, 1960s experimental feedback pieces by Robert Ashley and Gordon Mumma, among others). An effort to form such a corpus must be accompanied by efforts to annotate properly at least a representative subset, so that a musicological ground truth is established for guidance. Finally, there are many more computational techniques to explore, such as navigational tools through corpora, or the use of complexity analysis to measure the ability of data compression algorithms to treat noise music.

Eryck Abecassis

What is noise (music) to you?

I like to define noise with its mathematical and physical definitions.

In those domains, noises are all the non-desirable, distorted events in the normal signal.

Noise music is also a way to put an ear into unknown territories.

Why do you make it?

Because I always have, since the beginning of my musical practice; as long as I can remember, noise has been present.

Also I cannot imagine myself having a thought without envisaging its "disturbing counterpart."

Just for spirit sanity.

NOISE IN AND AS MUSIC

Inside Fama's House: listening, intimacy, and the noises of the body

Martin Iddon

1

Fama, the Roman embodiment of "rumor," enjoys a mixed reputation. Virgil complained that rumor was "the swiftest of all evils/*malum qua non aliud velocius ullum*,"[1] while Shakespeare was convinced that none could stop the "vent of hearing when loud Rumour speaks."[2] Even though Ovid claimed that rumor "loves to tangle true with false/*quae veris addere falsa gaudet*" and Shakespeare had Rumour herself proclaim "[u]pon my tongue continual slanders ride,"[3] in truth, for the most part it seems that the veracity of the reports of Fama are typically unquestionable. While the precise sort of fame she presages may be unwelcome—and this is certainly the "evil" of which Virgil speaks—that Fama is an accurate witness of what she has heard is hardly in question.[4] If there is slander in her words, it originates in the lies that she accurately and truthfully reports. The ways in which the Latin word, *fama*, intersects with related words is, too, suggestive. The Greek, *phēmē*, from which *fama* derives, does bespeak "rumor" or "gossip," but it is also an "utterance" or a "voice." The Latin *rumor* has, for a contemporary English-speaking audience, become intermingled with *fama*. The French *rumeur*, which derives from it, makes clear the elision; it has two meanings: an indeterminate noise and the

---

1 Publius Vergilius Maro, *Aeneid* IV, 74.

2 William Shakespeare, *Henry IV*, Part II, Prologue, 2.

3 Publius Ovidius Naso, *Metamorphoses* IX, 138–39; William Shakespeare, *Henry IV*, Part II, Prologue, 6.

4 See Nancy Zumwalt, "*Fama Subversa*: Theme and Structure in Ovid *Metamorphoses* 12," *California Studies in Classical Antiquity* (1977), vol. 10: 221, n. 11.

communication of a piece of gossip.⁵ *Fama*, then, might represent the sorts of noise *concealed* by the noises of rumor: quiet whispering, rustling murmurs, hidden by the louder noises which rumor has come to designate, as, perhaps most obviously, in Russolo's *L'Arte dei rumori* (1913).

In the first instance, however, my concern is with the site of listening. Importantly, as Due notes, while Virgil's and Shakespeare's conceptions of Rumour portray her as a horrific beast, abroad in the land, Ovid concentrates not on the creature Fama but, instead, on her dwelling place:⁶

> *Orbe locus medio est inter terrasque fretumque*
> *caelestesque plagas, triplicis confinia mundi;*
> *unde quod est usquam, quamvis regionibus absit,*
> *inspicitur, penetratque cavas vox omnis ad aures:*
> *Fama tenet summaque domum sibi legit in arce,*
> *innumerosque aditus ac mille foramina tectis*
> *addidit et nullis inclusit limina portis;*
> *nocte dieque patet: tota est ex aere sonanti,*
> *tota fremit vocesque refert iteratque quod audit;*
> *nulla quies intus nullaque silentia parte,*
> *nec tamen est clamor, sed parvae murmura vocis,*
> *qualia de pelagi, siquis procul audiat, undis*
> *esse solent, qualemve sonum, cum Iuppiter atras*
> *increpuit nubes, extrema tonitrua reddunt.*

---

5   A further, curious elision might be noted: the slang word scuttlebutt too refers to gossip (and is a possible meaning of the contemporary French *rumeur*), but is derived from a nautical term for the water butt on a ship; in contemporary usage, this has to do with the idea of rumors being spread around the workplace water cooler. The English word noise itself stems from the Latin *nausea*, which is to say seasickness, in turn derived from Greek ναῦς, a ship. Rumor and noise, in a linguistic sense, are profoundly interwoven concepts from the outset, even if the precise relationship between them is in constant flux. *Fama* too, it should be remembered, has also had the meaning of "reputation" in medieval Europe, though it was the "reputation" an individual had as constructed through what was said about him or her. See Thelma Fenster and Daniel Lord Smail (eds.), *Fama: The Politics of Talk & Reputation in Medieval Europe* (Ithaca: Cornell University Press, 2003).

6   See Otto Steen Due, *Changing Forms: Studies in the Metamorphoses of Ovid* (Copenhagen: Gyldendal, 1974), 148–49.

INSIDE FAMA'S HOUSE

At the world's center lies a place between
The lands and seas and regions of the sky,
The limits of the threefold universe,
Whence all things everywhere, however far,
Are scanned and watched, and every voice and word
Reaches its listening ears. Here Rumour dwells,
Her chosen home set on the highest peak,
Constructed with a thousand apertures
And countless entrances and never a door.
It's open night and day and built throughout
Of echoing bronze; it all reverberates,
Repeating voices, doubling what it hears.
Inside, no peace, no silence anywhere,
And yet no noise, but muted murmurings
Like waves one hears of some far-distant sea,
Or like a last late rumbling thunder roll.

Ovid, *Metamorphoses* XII: 39–52

Where, for Virgil, Fama is abroad in the world, spreading rumor wherever she passes, in Ovid's conception Fama is a receiver, hearing all. Ovid proposes that many dwell alongside Fama here: Credulitas (credulity), Error (error), Laetitia (delight), Susurri (whispers), Seditio (sedition), and Timores (fear). Against this essentially negative characterization of what happens in Fama's house, in what follows I intend to suggest something rather different. I propose that the site of listening—Fama's house itself—may be understood as a metaphorical description of the listening body and the sorts of noises that that body encounters *as* a listening body. In developing this essentially corporeal, visceral approach to listening, I simultaneously argue that listening is *essentially* an intimate act, an activity that is, at heart, private and bodily. As Michel de Certeau puts it, linking bodies and listening in direct terms: "Through the legends and phantoms whose audible citations continue to haunt everyday life, one can maintain a tradition of the body, which is heard but not seen."[7]

2

Reading Ovid's description of Fama's house, one might hear, at first blush, echoes of an obscure text by Leibniz in which a perfect political system is described through the allegory of the "Palace of Marvels":

> These buildings will be constructed in such a way that the master of the house will be able to hear and see everything that is said and done without himself being perceived, by means of mirrors and pipes, which will be a most important thing for the State, and a kind of political confessional.[8]

Yet hearing is, surely, first hearing oneself in the internal cavity of one's own skull, before it is hearing anything else, and the question of who is master of such a house is hardly a simple one. Living with the sounds of one's own voice, "Is that really how I sound?" one might enquire of a recording of it.

The body barely allows the possibility of *observing* oneself. The visual is concerned with looking at others. Yet the aural turns inward before it turns outward. Not only this but, as Wannenwetsch emphasizes, hearing precedes seeing: "In the womb, and in our first days on earth, we see only very indistinctly. We do not initially identify our mother visually at all; we get to know her first of all precisely when she talks to us."[9] Nancy goes further.

---

[7] Michel de Certeau, *The Practice of Everyday Life*, trans. Steven Rendall (Berkeley: University of California Press, 1984), 163.

[8] Gottfried Wilhelm Leibniz, quoted in Jacques Attali, *Noise: The Political Economy of Music*, trans. Brian Massumi (Minneapolis: University of Minnesota Press, 1985), 7.

[9] Bernd Wannenwetsch, "'Take Heed What Ye Hear': Listening as a Moral, Transcendental and Sacramental Act," *Journal of the Royal Musical Association*, vol. 135, Special Issue no.1 (2010): 96.

Asserting, with Wannenwetsch, that "the body was *conceived* in darkness," he observes that the body's uterine existence was "shaped in Plato's cave, as the cave itself: prison or tomb of the soul." Yet though Nancy suggests that the "cave-body is the space of the body *seeing itself from within*,"[10] surely Wannenwetsch is closer to the mark in his implicit suggestion that such a cave-body is the space of the self *hearing itself* from within.[11] It is in this sense that Serres is able to claim that music "built our house before we were born as speaking beings—and not only in the vibrating enclosure of the uterus—and paved the way for our collective existence; the social contract, hidden from all languages, can be heard indistinctly in its orchestration."[12] Like Serres, Nancy asserts the primacy of hearing in the womb over the later sense of sight, in asking "[w]hat is the belly of a pregnant woman, if not the space or the antrum where a new instrument comes to resound, a new *organon*, which comes to fold in on itself, then to move, receiving from outside only sounds, which, when the day comes, it will begin to echo through its cry?"[13] Yet, further, Nancy suggests that the house that was built before birth is inescapable. In listening, one is always in some sense still "in the cave": "it is always in the belly that we—man or woman—end up listening. The ear opens onto the sonorous cave that we then become."[14]

Serres's account of listening hinges on precisely the question of hearing the sound of one's own body: health comes when the noise of the body's

---

10 Jean-Luc Nancy, "Corpus," *Corpus*, trans. Richard A. Rand (New York: Fordham University Press, 2008), 67.

11 See Martin Iddon, "Plato's Chamber of Secrets: On eavesdropping and truth(s)," *Performance Research* (2010), 15, no.3: 6–10.

12 Michel Serres, *The Five Senses: A Philosophy of Mingled Bodies*, trans. Margaret Sankey and Peter Cowley (London: Continuum, 2008), 123.

13 Jean-Luc Nancy, *Listening*, trans. Charlotte Mandell (New York: Fordham University Press, 2007), 37.

14 Nancy, *Listening*, 37.

organs is silenced.¹⁵ Serres's account is—hardly surprising in the context of a description of healing at Epidauros—reliant upon an ancient tradition of thought: "[i]t is a commonplace of ancient philosophical writing that internal uproar, the noise of the passions or of a mind divided against itself, prevents us from listening properly to the healing words of philosophy."¹⁶

For Nancy, almost the opposite is the case. Hearing, the hearing of oneself in particular, is at the origin of subjectivity:

> Perhaps we should thus understand the child who is born with his first cry as himself being—his being or his subjectivity—the sudden expansion of an echo chamber, a vault where what tears him away and what summons him resound at once, setting in vibration a column of air, of flesh, which sounds at its apertures: body and soul of some *one* new and unique. Someone who comes to himself by hearing him*self* cry (answering the other? calling him?), or sing, always each tome, beneath each word, crying or singing, *exclaiming* as he did by coming into the world."¹⁷

It is not a question of some return to a mythical time of full wholeness that Nancy proposes, rather more that listening is always a memory or an echo of "the resonance of being, or … being as resonance." Silence, in this case, would be neither privation nor the sound of health, as Serres has it, but "an arrangement of resonance," as when, recalling Cage's experience in the anechoic chamber, "in a perfect condition of silence you hear your own body resonate, your own breath, your heart and all its resounding cave."¹⁸ In this, at

---

15   Serres, *The Five Senses*, 85.

16   William Fitzgerald, "Listening, Ancient and Modern," *Journal of the Royal Musical Association*, 135, Special Issue no.1 (2010): 33.

17   Nancy, *Listening*, 17–18.

18   Nancy, *Listening*, 21.

least, Nancy and Serres agree: an escape from the cave, from the chamber of listening, is not possible, even if it were desirable:

> No matter how far I travel, poor subject that I am, I never manage to put any distance between myself and the droning of the language that shaped me. What merely resonated within my mother's womb is a clamour in this stone conch, and finds itself echoed in my innermost ear.[19]

Not for nothing does Serres emphasize the link to a shell, not only a protective casement for the body but also that potent metaphor for the ear in both shape and function. The rumors that Ovid hears are, finally, like the waves "of some far-distant sea." The shell placed against the ear—literally, as well as metaphorically, shell-to-shell—the body hears itself. As Bachelard neatly summarizes it, "[w]e might say that the inside of a man's body is an assemblage of shells."[20] Bachelard states, too, that "[e]verything about a creature that comes out of a shell is dialectical. And since it does not come out entirely, the part that comes out contradicts the part that remains inside. The creature's rear parts remain imprisoned in the solid geometric forms."[21] The dichotomy of inside and outside is surely not literally meant. Here, once the link between shells and ears has been made, the return to the transcendent Platonic world of pure forms can be seen: if the part within the shell, within the womb, or within the cave—however one conceives it—is in contradiction with the outer part—that which *looks* on the world—it is in listening that some approach to the transcendent experience of the Platonic real might, perhaps, be found.

---

19   Serres, *The Five Senses*, 93.

20   Gaston Bachelard, *The Poetics of Space*, trans. Maria Jolas (Boston: Beacon, 1994), 113.

21   Bachelard, *The Poetics of Space*, 108.

3

That listening is always, first and foremost, a listening to the self, if not *of* the self, is suggested by René Daumal:

> *Ecoute bien pourtant. Non pas mes paroles, mais le tumulte qui s'elève en ton corps lorsque tu t'écoutes. Ce sont des rumeurs de combat, des ronflements de dormeur, des cris de bêtes, le bruit de tout un univers.*

> Listen well. Not to my words, but to the noises that build up inside your body when you listen to yourself. They are the rumors of combat, the snores of sleepers, the cries of animals, the noise of an entire universe.[22]

Elsewhere, Daumal stresses the point, with a combination of imagery of particular relevance here:

> [...] *peau pleine de rumeurs aux échos de villes souterraines* [...]

> [...] skin full of rumors from the echoes of subterranean cities [...].[23]

The poem from which this second quotation is drawn is entitled "A perdre sens": to lose one's way, to be sure, but also to lose one's sense or, for that matter, meaning. Nancy plays on the multiple meanings of *sens* throughout his account of listening.

As he puts it, "sensing [*sentir*] (*aisthesis*) is always a perception [*ressentir*], that is, a feeling-oneself-feel [*se-sentir-sentir*]: or, if you prefer, sensing is a subject, or it does not sense."[24] The sensing subject is, in Nancy's terms, the

---

22 René Daumal, *Le Contre-ciel suivi de Les dernières Paroles du poète,* trans. Kelton W. Knight (New York: Tusk Ivories, 1990), 54. Translation modified.

23 Daumal, *Le Contre-ciel*, 118. Translation modified.

24 Nancy, *Listening*, 8.

only subject. Without sensing, there is no subject, no self. Those subterranean cities are redolent, too, of Serres's description of hearing the sound of one's own voice:

> We can neither speak nor sing without the feedback loop which guarantees the audibility of our own voice. The ear guarantees and regulates the mouth, which emits noise in part for the speaker, in part for others, who in turn guarantee other feedback loops. Intuitively we imagine a large prostrate body, buried underground, the marble pavilion of its ear jutting out, its dark mouths speaking and shouting for millennia through the plunging cliff-face.[25]

Significant, too, is Nancy's re-iteration that "the ears don't have eyelids." No more does Fama's house have doors. The sound cannot, or cannot simply, be shut out. As Bull observes, in the case of Odysseus's encounter with the Sirens: "[i]t is not the seeing or touching of the Sirens that motivates Odysseus but the hearing of their song; it literally enters him."[26] Just the same sort of penetration, or opening up, is at play in de Certeau's description of the voices of bodies—though here perhaps it is a question of a reclamation, a return rather than an echo—in which psychoanalysis moves from "a 'science of dreams' to the experience of what speaking voices change in the dark grotto of the bodies that hear them." As de Certeau has it, this chamber—grotto or cave—becomes a "plural body in which ephemeral rumors circulate."[27] Daumal's subterranean city is, too, a body. The cave, Fama's house, and the body become one and the same subject here in the juxtaposition of Daumal, Nancy, and Serres.

---

25 Serres, *The Five Senses*, 110.

26 Michael Bull, "Thinking about Sound, Proximity, and Distance in Western Experience: The Case of Odysseus's Walkman," in *Hearing Cultures: Essays on Sound, Listening and Modernity*, ed. Veit Erlmann (Oxford: Berg, 2004), 178.

27 de Certeau, *Practice of Everyday Life*, 162.

It is in listening, then, that this is most potent, in the case of the resonant chamber of the body, which already experiences the distance of an echo [*renvoi*]. To listen is to be penetrated by sound in Nancy's terms and, more significantly, to be *opened up* by sound: "it opens me inside me as well as outside, and it is through such a double, quadruple, or sextuple opening that a 'self' can take place."[28] The openings of Nancy's listening might be seen as little different from the countless entrances of Fama's house, the body itself.

It might be necessary in such a context to revise that notion that suggests that listening tilts at pure internality, pure intimacy of the self, since "[t]o be listening is to be *at the same time* outside and inside, to be open *from* without and *from* within, hence from one to the other and from one in the other."[29] Even the statement that hearing is first the hearing of oneself must be revised, since listening in such terms then becomes the hearing of others *within* oneself, resounding as one hears oneself too. Listening becomes a "sharing of an inside/outside, division and participation, de-connection and contagion."[30] It is notable that the intermingling of sound, listening, and illness that pervades Serres's account is retained by Nancy too, even if in modified form.

This double listening is emphasized by Wannenwetsch: "[I]n listening to a piece of music, we listen not only to the musicians, but also to the sound of our own body resonance, as a way of responding to the spirit of the piece and its truth that claims and seeks to transform us."[31] The resonance of the

---

28 Nancy, *Listening*, 14.

29 Nancy, *Listening*, 14.

30 Nancy, *Listening*, 14.

31 Wannenwetsch, "Take Heed What Ye Hear," 102. No surprise, then, that in Wannenwetsch's essentially theological account, the Word is made flesh through the bodily resonance of song: "If psalmody does not engage scripture as script, but scripture as voice—listening to a word that cannot be 'interpreted' but only 'incorporated': when the human body—as the chamber of resonance of the sounding scripture—takes on scripture's own vibrating frequency, so to speak" (p. 100). Listening to the resonance of the self is, in his terms, another kind of truth-unveiling. As Wannenwetsch conceives it, this is, in part at least, what it would mean to be "all ears," "perceptive and responsive to God's address and the story of his ways with mankind. That is why the 'heart' is given prominence in the *Shema* as the 'organ' in which all sensual perception, thought and will are given direction." (p. 93).

self, in the act of listening, not only opens the body up to the self, it changes the self as it listens *to itself*. The sonorous place is not, as Nancy stresses, "a place where the subject comes to make himself heard,"[32] even if the first echo is that of one's own voice and, furthermore, even if "the sound that penetrates through the ear propagates throughout the entire body something of its effects, which could not be said to occur in the same way with the visual signal."[33] Rather "it is a place that becomes subject insofar as sound resonates there." The reverberation chamber of listening becomes, in such a construction, "nothing other than the body from end to end." Meaning [*sens*] is already in a state of return before it has acquired meaningfulness, "completely ahead of signification, meaning in its nascent state, in the state of return [*renvoi*] for which the end of this return is not given (the concept, the idea, the information)." Far from being in any sort of progress toward meaningfulness, the listening body's sense is confined by "the state of return without end, like an echo that continues on its own and that *is* nothing but this continuance going in a *decrescendo*, or even in *moriendo*."[34]

Nancy's version of the eternal return of the self here is, itself, echoed in Ashbery's words, which themselves recall the central metaphorical descriptions of the body at play here: it is rock, from which a cave might be carved by the water. His description of a wave is obviously pertinent in any consideration of the wave that sound cannot but be: "Like a wave breaking on a rock, giving up/Its shape in a gesture which expresses that shape."[35] This need not be the self-annihilation of the work of art or the transience of the subject as such, but only, as Nancy's thinking suggests, the gradual echoing

---

32 Nancy, *Listening*, 17.

33 Nancy, *Listening*, 14–15.

34 Nancy, *Listening*, 27.

35 John Ashbery, "Self-Portrait in a Convex Mirror," *Self-Portrait in a Convex Mirror* (Manchester: Carcanet, 1985), 73.

moriendo of the self in the body.³⁶ Indeed, Nancy comes close to making the same argument. Situating listening as the sense of which other senses are, in a certain context, themselves echoes, he remodulates, once again, his central theme: "*Sense*, here, is the ricochet, the repercussion, the reverberation: the echo in a given body, even *as* this given body, or even as the gift to *self* of this given body."³⁷ Nancy recalls Wittgenstein's imagined experience of listening to a sound disassociated from its timbre, and that Wittgenstein took timbre as a particular instantiation of his notion of private experience (which is to say, in Nancy's reading, experience which is not communicable).³⁸ Yet Nancy's conclusions stray some way from Wittgenstein's. Unwilling to brook the incommunicable as such, Nancy suggests instead that "timbre is communication of the incommunicable: provided it is understood that the incommunicable is nothing other, in a perfectly logical way, than communication itself, that thing by which a subject makes an echo—of self, of the other, it's all one—it's all one in the plural."³⁹ Communication, listening, sense, and meaning all resound inside and between bodies and, if Nancy's underlying notion that the self is constituted in this eternal returning echo, being in his terms is neither singular nor plural, neither outside nor inside. Nevertheless, it *is* both bodily and fundamentally intimate.

4

De Certeau hits on precisely the intersection between enunciation, listening and intimacy, recalling in his description both Wittgensteinian

---

36 See, for instance, Michael E. Hattersley, *Socrates and Jesus: the Argument That Shaped Western Civilisation* (New York, NY: Algora, 2009), 189; or, Jody Norton, *Narcissus sous rature: Male Subjectivity in Contemporary American Poetry* (Cranbury: Associated University Presses, 2000), 203.

37 Nancy, *Listening*, 40.

38 See Ludwig Wittgenstein, "Notes for Lectures on 'Private Experience' and 'Sense Data,'" *Philosophical Occasions, 1912–1951* (Indianapolis: Hackett, 1993), 200–88.

39 Nancy, *Listening*, 41.

incommunicability and the urgency of echoing exchange that is central to Nancy's description:

> "There are" everywhere such resonances produced by the body when it is touched, like "moans" and sounds of love, cries breaking open the text that they make proliferate around them, enunciative gaps in a syntagmatic organization of statements. They are the linguistic analogues of an erection, or of a nameless pain, or of tears: voices without language, enunciations flowing from the remembering and opaque body when it no longer has the space that the voice of the other offers for amorous or indebted speech. Cries and tears: an aphasic enunciation of what appears without one's knowing where it came from (from what obscures debt or writing of the body) without one's knowing how it could be said except through the other's voice.[40]

The key relationships between the ideas presented thus far are intertwined in de Certeau's account. He says, first, that "[t]hese contextless voice-gaps, these 'obscene' citations of bodies, these sounds waiting for a language, seem to certify, by a 'disorder' secretly referred to an unknown order, that there is something else, something other." The echo of the impossibility of communication is evoked again here, but, more potently, he continues: "they narrate interminably (it goes on murmuring endlessly) the expectation of an impossible presence that transforms into its own body the traces it has left behind."[41] The ceaseless murmur (this, too, a trace of the Latin *rumor*) echoes into selfhood, even without wholly fulfilling the promise of full presence. Through the penetration of the body by rumor, the body, resounding, is transformed. That desire for the resonance to begin to signify a birth to full presence is highlighted by Serres, too, in a similar context:

---

40   de Certeau, *Practice of Everyday Life*, 163.

41   de Certeau, *Practice of Everyday Life*, 163–64.

> Global, integral, already abstract hearing, seeking unity, fills volumes: boxes, cases, houses, prisons, theatres, cities, circuses, hells and forests, marine expanses on which the musician's severed head, forgotten, detached, drifts on towards distant islands and sings, pervading the wind that sweeps between sky and waves. And my whole body, a music or language box, resonance chamber, resounding gong; and my local group, an assembly sometimes found in the theatre, for contemplation.[42]

In a similar list, a little later in his account, Serres emphasizes that the resonant chamber of the body is also an intimate space, a home for the self:

> I am the home and hearth of sound, hearing and voice all in one, black box and echo, hammer and anvil, echo chamber, music cassette, pavilion, question mark drifting through the space of meaningful or meaningless messages, emerging from my own shell or drowning in the sound waves, I am nothing but empty space and a musical note, I am empty space and note combined.[43]

Again, the reiteration is of bodies, caves, and shells—all boxes as Serres insists—but also the homely site of the self. He demands that the bodily experience of sound remains poised between meaningfulness and meaninglessness, both empty and full. Such a positing of the body as resonant home recalls Fama's house, of course, but also points toward the way in which Bachelard conceives

---

42 Serres, *The Five Senses*, 138. The severed musician's head mentioned by Serres here refers to the myth that, even after his death, Orpheus's head continued to sing as it was washed down the Hebrus. The death of Orpheus is recounted by Ovid in *Metamorphoses* XI: 1–60. The image of Orpheus's severed head was a relatively popular nineteenth-century image in painting, featuring in work by Gustave Moreau, Jean Delville, and John William Waterhouse. The head of Orpheus also plays a central part in Russell Hoban's novel, *The Medusa Frequency* (London: Cape, 1987).

43 Serres, *The Five Senses*, 141.

of the poetics of the home, an aesthetics which might prove fruitful when extended to its bodily analogue.[44]

Bachelard certainly makes the necessary links, if implicitly. His description of the house conjoins it directly with the self, not least since his concern is with the intersection of space and psychoanalytic theory:

> thanks to the house, a great many of our memories are housed, and if the house is a bit elaborate, if it has a cellar and a garret, nooks and corridors, our memories have refuges that are all the more clearly delineated. All our lives we come back to them in our daydreams. A psychoanalyst should, therefore, turn his attention to this simple localization of our memories. I should like to give the name of topoanalysis to this auxiliary of psychoanalysis. Topoanalysis, then, would be the systematic psychological study of the sites of our intimate lives.[45]

His statement that "a knowledge of intimacy, localization in the spaces of our intimacy is more urgent than determination of dates" also recalls more recent discussion of sound and space, in just this mode of constructing the intimate space of the home while abroad in the world.[46] Michael Bull, for one, extrapolates much from Berkeley's note that "bodies and external things are not properly the object of hearing; but only sounds, by the mediation whereof the idea of this or that body or distance is suggested to

---

44 It should be noted that Bachelard himself is leery of the use of image as if it were metaphor, observing that Bergson's use of the word drawer is almost always as a disdainful metaphor. This seems to Bachelard to be "a good example for demonstrating the radical difference between image and metaphor. I shall therefore insist upon this difference before returning to my examination of the images of intimacy that are in harmony with drawers and chests, as also with all the other hiding-places in which human beings, great dreamers of locks, keep or hide their secrets" (Bachelard, *Poetics of Space*, 74). I do not, then, intend to suggest that Bachelard would necessarily have sympathized with the way in which I deploy his thought here.

45 Bachelard, *Poetics of Space*, 8.

46 Bachelard, *Poetics of Space*, 9.

his thoughts."[47] Jonathan Rée's account claims, following this, that hearing should be understood as a contact sense and that, more pertinently here, "[s]ounds, to use a phrase of Berkeley's, are perhaps 'as near to us as our own thoughts.'"[48] It is from this matrix that Bull identifies an absence of the aural in considerations of space and distance.[49] Raymond Williams, Bull stresses, would be likely to have found nothing concerning in the portable, intimate homely spaces of (then) Walkmans or (now) iPods: "It is not living in a cut off way, not in a shell that is just stuck. It is a shell you can take with you, which you can fly to places that previous generations could never imagine visiting."[50] Adorno, by contrast, "never succumbed to the temptation to split off spheres of experience in his analysis of Western consumer culture; for him the experiences of the street and the spaces of the home were always intimately linked."[51] Such listening activities, the taking with one of one's own portable shell, are

> about the desire for proximity, for a mediated presence that shrinks space into something manageable and habitable. Sound, more than any other sense, appears to perform a largely utopian function in this desire for proximity and connectedness. Mediated sound reproduction enables

---

47 George Berkeley, *An Essay Towards a New Theory of Vision* (Cirencester: Echo Library, 2005 [1709]), 19.

48 Jonathan Rée, *I See a Voice: A Philosophical History of Language, Deafness and the Senses* (London: Flamingo, 2000), 35–36. Rée also notes that "it is not difficult to make something erotic of the idea that sounds have to enter into our bodies before they can be heard" (p. 35). Something of the same sense can clearly be felt in Nancy's account of a penetration and opening up of the body by sound.

49 It should be noted that, since Bull's account, the ground has shifted at least a little with the publication of volumes such as Barry Blesser and Linda-Ruth Salter's *Spaces Speak, Are You Listening? Experiencing Aural Architecture* (Cambridge: The MIT Press, 2006).

50 Raymond Williams, *Television: Technology and Cultural Form* (New York: Routledge, 2003), 161; quoted in Bull, "Thinking about Sound, Proximity, and Distance," 174.

51 Bull, "Thinking about Sound, Proximity, and Distance," 176.

consumers to create intimate, manageable, and aestheticized spaces in which they are increasingly able to, and desire to, live.[52]

This sort of activity, then, is properly nostalgic, whatever its technological foundations. Perhaps in simulated form at least, this approach to listening nears an answer to the question Bachelard asks, even if only in part:

> Transcending our memories of all the houses in which we have found shelter, above and beyond all the houses we have dreamed we lived in, can we isolate an intimate, concrete essence that would be a justification of the uncommon value of all of our images of protected intimacy?[53]

Even so, Bachelard's description is not without a particular desire for the, in this context, almost wholly unobtainable. He asks, too: "How …, in these fragments of space, did the human being achieve silence? How did he relish the very special silence of the various retreats of solitary daydreaming?"[54] Bachelard's response echoes, at one remove, Adorno's. Bachelard's critique of a certain philosophical position—that of withdrawal into 'pure' thought—takes on an added resonance in this case:

> Outside and inside are both intimate—they are always ready to be reversed, to exchange their hostility. … Intimate space loses its clarity, while exterior space loses its void, void being the raw material of possibility of being. We are banished from the realm of possibility.
>
> In this drama of intimate geometry, where should one live? The philosopher's advice to withdraw into oneself in order to take one's place in existence, loses its value, and even its significance, when the supplest

---

52 Bull, "Thinking about Sound, Proximity, and Distance," 177.

53 Bachelard, *Poetics of Space*, 3.

54 Bachelard, *Poetics of Space*, 9.

image of 'being-there' has just been experienced through the ontological nightmare of this poet [Henri Michaux].⁵⁵

No surprise, in the light of what has been outlined thus far, that Michaux's account is precisely to do with sound. Indeed, it is not merely sound that concerns Michaux, but something very precise: the long rumbling of distant thunder, which Ovid paired with the distant crashing of the sea:

L'espace, mais vous ne pouvez concevoir, cet horrible en dedans—en dehors qu'est le vrai espace.
Certaines (ombres) surtout se bandant une dernière fois, font un effort désespéré pour 'être dans leur seule unité.' Mal leur en prend. J'en rencontrai une.
Détruite par châtiment, elle n'était plus qu'un bruit, mais énorme.
Un monde immense l'entendait encore, mais elle n'était plus, devenue seulement et uniquement un bruit, qui allait rouler encore des siècles mais destiné à s'éteindre complètement, comme si elle n'avait jamais été.

Space, but you cannot even conceive the horrible inside-outside that real space is.
Certain (shades) especially, girding their loins one last time, make a desperate effort to "exist as a single unity." But they rue the day. I met one of them.
Destroyed by punishment, it was reduced to a noise, a thunderous noise.
An immense world still heard it, but it no longer existed, having become simply and solely a noise, which was to rumble on for centuries longer, but was fated to die out *completely*, as though it had never existed.⁵⁶

---

55 Bachelard, *Poetics of Space*, 217–18.

56 Henri Michaux, *Nouvelles de l'étranger* (Paris: Mercure de France, 1952); quoted in Bachelard, *Poetics of Space*, 216–17. (Translation as it appears in Bachelard.)

Bachelard, through Michaux, ties together many of the fundamental concerns of the above. Bachelard reads in Michaux a description of "a spirit that has lost its 'being-there' [*être-là*], one that has so declined as to fall from *the being of its shade* and mingle with the rumors of being, in the form of meaningless noise, of a confused hum that *cannot be located*." Yet Bachelard is unconvinced by this seeming fall from grace. He argues that the spirit *always was* what it is in Michaux's prose poem: "a sonorous echo." As Bachelard concludes, being in the world, being the resonant chamber of the self that is constituted in selfhood only by the recurring echoing of the self itself, "[t]he hum of the being of rumors continues both in time and in space."[57]

---

57 Bachelard, *Poetics of Space*, 217.

Part 2: Practices

George Lewis

What is noise (music) to you? Why do you make it?

Come to think of it, looking back on the past 40 years or so, I certainly have made a lot of what many people would call "noise—just noise." In fact, I've often said that I want my music to be like the weather, a fractal, self-similar noise that both poses and responds to conditions. I've also been associated with a diverse range of noisy people from around the world—the Great Chain of Noisy Beings.

So why do I do it? First of all, of course anyone can make noise, and noisiness is actually expected from some of us. But your question, with its parenthetical invocation of "music," seems to imply some great division between noise and music, or maybe even a bit of a hierarchy that moves from music to sound, and finally to noise. Because of its supposed extreme transgression (as distinct from the mere ineffability of music), noise becomes the highest stage of sonic utterance, and therefore the most desirable.

Certainly, many conceptions of noise tend to embed desire. We want something from our noise—we don't want it to be just sound—and when people announce themselves as noisers, we want something from them too.

My experience of that desire dovetails with the connection between noise, improvisation, spontaneity, and nomadism. As Gilles Deleuze and Felix Guattari observe in *A Thousand Plateaus*, "[i]t is a vital concern of every State not only to vanquish nomadism but to control migrations and more generally, to establish a zone of rights over an entire 'exterior'. ... If it can help it, the State does not dissociate itself from a process of capture of flows of all kinds, populations, commodities or commerce, money or capital, etc."

To these kinds of actants, tight control over the flow of noise is vital. In fact, for traditional authority, spontaneous action itself is noise. But in his 1999 book *Culture on the Margins,* historian Jon Cruz points out, the trickster function of noise as "sound out of order. It evades, eludes, spills out of, or flows over, the preferred channels—out of place, resistant to capture." In that sense, the pretense to control becomes exposed as quixotic; noise and noisers routinely overflow the banks of propriety, resisting and unleashing. People hear the sound and say, "no one told me it could be like that; I wonder what else they haven't told me." Or they say, "wow, that music is really different"; once they start down that road, thoughts inevitably turn to what else might need to be different.

So when we want change, in the memorable phrase of the rap group Public Enemy, we "bring the noise"—in Egypt, Tunisia, Montreal, or elsewhere. The improvised, spontaneous, seemingly leaderless nature of these and other protests reminds us of the primary remit of new music and new noises: to declare that change *is* possible.

And let's not forget all that freedom talk: noise symbolizes freedom, and noisers use noise to free themselves and others. Of course, noise symbolizes not the object, but the subject. Subalternity speaks in noisy cadences.

Sometimes we don't bring the noise ourselves, but eagerly await its arrival. Those who wait are hardly passive, but form part of the ever-shifting, self-organizing assemblage of noisy desire. If we stop making noise, other noisers become discouraged, and we can imagine that discouragement working to the advantage of entrenched interests whose primary remit is eliminating our consciousness of possibility.

So, as Michel Foucault put it in a late essay, "What Is Enlightenment?," we must "transform the critique conducted in the form of necessary limitation into a practical critique that takes the form of a possible crossing-over." Maybe a while back it was enough just to get the noises out, but nowadays I find that the really good noisers pay close attention to sonic ecology. Even though, early on, the sonic epoché ("letting sounds be themselves") worked as well for such budding artist-noisers as John Cage as Husserl's phenomenological one did for thinking about sound in time, not attending to the symbols of the local soundscape turns out to be a good way to fall behind the Darwinian curve—say, if you live in one of those 'hoods where you have to know when to hit the deck.

Also, as 21st-century people, we routinely encounter noise designed to throw us off the scent of change—what Anthony Braxton, writing his *Tri-Axium Writings* in Paris in the wake of Guy Debord, calls "spectacle-diversion." So part of what I do as a noiser involves training people to differentiate the sounds that empower from those that hamper and misdirect.

I once wrote that early African American free jazz was characterized by contemporaneous reception as embodying a tripartite assemblage consisting of anger, noise, and failure. Everybody knows that failure is an orphan, but it occurs to me now that very little self-characterized art music confronts anger, perhaps because the artists are afraid to say what they might be angry about; we've allowed you extra resources and your own bourgeois playpen, so they tell us.

But a sonic practice that embraces noise too insouciantly runs the risk of succumbing to the regulative force of genre, and thereby losing its noisiness. The afterimage of past transgressions can still be useful in reaping art world rewards, but we want an experimental music that can own up to the consequences of its noise, that can face up, when needed, to the possibility that there really is something to be angry about. In that way, noise and the noisers who make it can evade capture and live to noise again.

NOISE IN AND AS MUSIC

"We Need You To Play Some Music"

Phil Julian

At some point in the late 1980s I began to take an interest in "Noise Music." Noise is obviously a broad, subjective term, but by way of a personal definition this was music that made deliberate use of sounds generally regarded as non-musical. Anything unpleasant, awkward, or difficult to listen to, set within a traditional musical context: a record, a concert, and so on. The objectionable presented as entertainment.

The route to "Noise" for me was largely via 1960s and 70s heavy rock albums that could be found easily and cheaply in second-hand record shops. Commercially available Top 40 music in the late 1980s by and large lacked the visceral qualities inherent in a heavily amplified, guitar-based record.

What became key was looking for the extremes within these recordings: the strangest, fastest, slowest, loudest examples. It was not long before the most interesting parts of these records had nothing to do with the songs or the musicianship of the groups, but were instead things like the extended equipment demolitions and amp smashes that occasionally acted as a finale. Most of these were nothing more than over-the-top showmanship, but the sounds created via feedback and broken equipment became far more interesting to me than the half an hour of songs you had to wade through to get to them.

It was clear that what interested me was music that "overwhelms" the listener in some way, or more specifically that completely inhabits the space it is given. This does not necessarily have to be achieved via extremes of volume, but can come more via a physical "presence" to the sound.

Rock acts in the 1960s and 70s often made references in interviews and on record sleeve notes to other musicians who had been influential, and this often included avant-garde, jazz, academic, or classical musicians. This provided further spurs to investigation in the search for extreme sounds.

All of the records I discovered around this time acted as a set of permissions. Yes, you could have an album with just one song on it. You could play this loudly or that quietly within one piece of music. You could drag that furniture around, record it, and present it as you would a string quartet. It didn't take much excavation to realize you could do absolutely anything you wanted to with sound.

This is certainly where an important and clear distinction between music and sound became apparent to me. The most interesting of these avant/experimental/whatever records were not concerned with traditional musicality at all, and many had apparently little or no interest in even representing a musical experience. They didn't use formal conventions like songs or obvious narrative structure to prop up the "funny noises." They seemingly stripped away the unnecessary to focus far more on audio-related sensations and phenomena not usually directly associated with music—chance, error, silence, noise, repetition, and stasis—forcing the ear to create its own apophenic patterns.

At some point in the early 1990s I started to make my own recordings. Pre-internet, the initial inspiration came largely from cassettes and fanzine reports of various "junk noise" performers in Japan and elsewhere. Equipment could be minimal, self-built, and recycled; everyday objects along with salvaged electronics, microphones, and a half-working reel-to-reel tape machine were used to make much of the early material. The key component was not musical ability, access to expensive equipment, or even much of a technical grasp, but imagination. Having a restricted amount of equipment to work with requires much more in terms of creative thinking to achieve the desired results, but also (in hindsight) provides a focused framework.

At roughly the same time, laptop computers had started to be used in music production and for live performance. They were slow by modern standards, expensive, and unreliable, but in the right hands could generate fascinating sounds difficult to achieve using analog equipment. Toward the end of the 1990s, they began to appear more frequently as solo performance instruments. There was a strange, remote, austere quality to these live laptop performances that was intriguing to watch. Audience members and sound technicians in particular were often openly hostile toward performers, and, though now commonplace, there is still to this day a suspicion that a solo laptop performance is a cop-out or a cheat in some way. The title of this chapter, for example, comes from a particularly irate bar manager who felt the need to interrupt a live laptop performance mid-set to complain about the perceived lack of music I was producing. Sadly for her, I don't do requests.

Like most instruments, the computer is as complicated or as easy as one chooses to make it. A performer can certainly just hit the spacebar and play back something pre-recorded. But they can also use a computer to create live music in a completely non-prepared, improvised way. Unfortunately, both of these approaches look about the same to an audience: a person gazing at a small screen. This seemingly represents "poor ticket value" to an audience that still wants a rock and roll show or event of some kind—a common but perhaps slightly strange expectation from an experimental music event.

Noise, in theory, should be the ultimate arena for an "anything goes" mentality from audiences and performers alike. If the Punk ethos was "learn three chords and form a band," Noise effectively did away with chords and bands completely. By the time I was involved, Noise Music had been around long enough to develop stylistic moves and a feel of the "traditional" about it. The dynamics of live Noise events were often similar. The way the music was constructed and shaped had a tendency to follow certain quiet-to-loud patterns. Computers were still viewed with suspicion by audience and performers alike, but the sounds coming from them also had certain elements that were identifiable as common currency. In a small way, Noise had started to develop a fingerprint like any other genre of music that has been around for long enough. The feeling that a certain "sameness" had crept into the genre actually became a useful thing on a personal level. Things that had not been considered before became interesting, like structure in terms of volume and duration. Performer position (out of view of the audience, sitting in the audience rather than onstage, starting and finishing at unexpected moments, not being present at the performance at all, anonymity, etc.) became relevant. Ultimately it was important that these where not overused as they had a danger of becoming "tricks" that an audience would get used to seeing or expect in some way. In the end, the most relevant thing personally was to avoid obvious sound palettes and keep the instrumentation as broad as possible. Not being a musician means that all sounds have a potential use, whether they are derived from traditional instruments or elsewhere.

Fortunately, some of these slightly overused moves in Noise seem to have given way recently to a freer approach, and a general feeling of "anything goes" has started to emerge again. Because the term means different things to different people, Noise as a scene or genre can only ever be fragmentary and constantly evolving, something that leaves it in a unique, fluid position.

Lasse Marhaug

What is noise (music) to you?

Noise music is, simply said, a field of music in which elements that feature prominently in other styles of music—such as melody, rhythm, and tonal harmony—take a back seat. Noise music is about the sound itself, and how you structure that sound is what defines it as music.

Why do you make it?

Because I like it. I enjoy the sound of dense electronic overload, feedback, and distortion. I like how noise both offers a space to move around freely within, and a feeling that engulfs you. It pleases me both emotionally and intellectually. In performance noise is overwhelming and you can't get away from it. The music is often physical. You feel it. Noise offers freedom from the dictatorship of emotions found in traditional music. You get from it what you bring to it.

NOISE IN AND AS MUSIC

Beyond Pitch Organization: an interview with Michael Maierhof

Sebastian Berweck

Sebastian Berweck: Michael, I always thought of your work as a composer as that of a sound researcher, as somebody who tries to get new sounds out of objects and instruments. However, I did not see you in the line of *Merzbow* or early *Einstürzende Neubauten* and was therefore quite surprised when you labeled your music as "noise music." Therefore, my first question is a very general one: what constitutes noise music in your view and where do you see yourself in this context?

Michael Maierhof: I make a distinction between pitch-organized music and non-pitch-organized music. If one does not use pitches in music as a central category, people often describe this as *Geräusch-Musik*, which would be the German equivalent to noise music. And because I am not writing pitch-organized music, my music belongs to the other category, which is usually called noise.

However, my strategy is to define a more elaborate concept of noise, something that I call *Klangkomplexe*, to describe the complex qualities of non-pitch materials.

SB: Can you explain how you make these sounds? What tools do you use and how do you find them?

MM: Because I mostly work with acoustic instruments, let us exclude the possibilities of electronics for the moment.

One possibility is to create new instruments that can produce the sound complexes I am looking for. Since 1999 I have built *resonance boxes* with different structured surfaces, which can be found on the backside of tiles, on glass, on metal gratings, and acrylic glass, for example. The performers play on these structures by using different tools to set these surfaces into vibration, for example by drawing circles with glass balls on them.

These surfaces are often structured, and when the performers draw a simple circle over the structure they can easily produce quite intricate sounds. This is a method I use, for example, in my percussion pieces *Untergrund 1* for five percussion players, *Untergrund 2* for four percussion players, and *splitting 19.1* and *splitting 24*, which are for percussion solo. See, for example, *splitting 19.1* (Figure 1), where the percussion player performs different figures with objects like glass marbles on a set of different metal gratings, which are separated and muted by gaffer tape. In measure 50 the right hand moves over ten different sections, each of which has a different sound.[1] This single movement produces a very complex sound within the timeframe of only one beat. The sound can then be analyzed and described according to the sound characteristics of each section, their specific combination, and their density in time.

The same principle can be observed in *splitting 24* (Figure 2). Here, performers move sets of glass marbles, which act as multiphonic activators, on different acrylic glass structures.[2]

The second possibility is to use traditional instruments—mostly pitch instruments—and redefine them through *preparations*, *applications*, and *new activators*.

Preparations are mostly used for transforming and/or disturbing the physical process of the instrument's vibration and thereby splitting the originally single-pitched sound into multiphonics. The model for this is, of course, John Cage's preparation of the grand piano strings with screws.

---

1 White regions in the diagrams above the staff indicate sections, which are separated by the gray gaffer-tape fields.

2 Indicated by gray and white regions in the figures above the staff (editor's note).

BEYOND PITCH ORGANIZATION

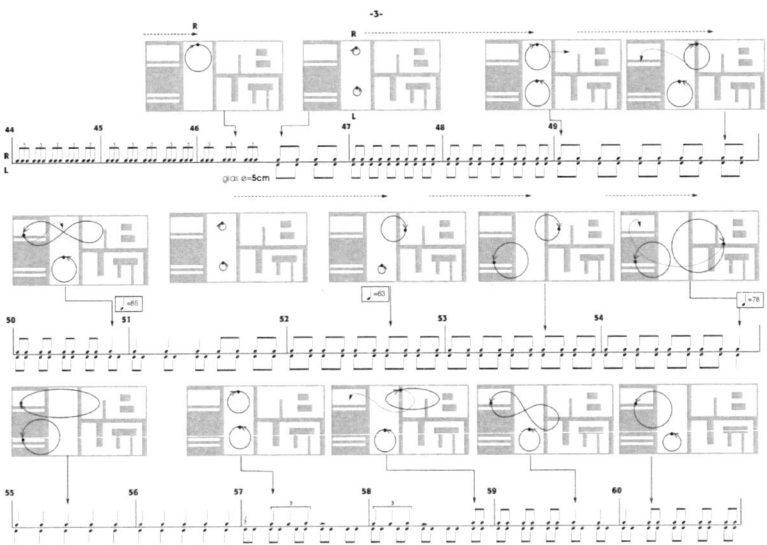

Figure 1: *splitting 19.1*, page 3

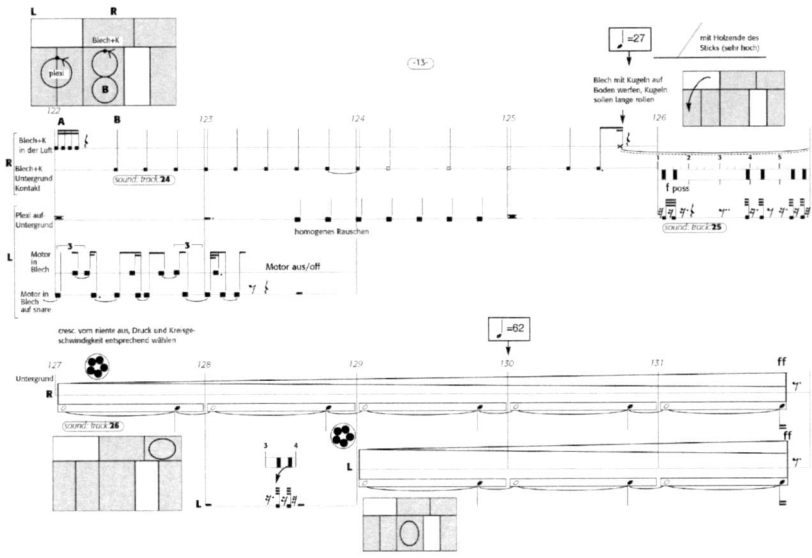

Figure 2: *splitting 24*, page 13

Since the nineties I have used, for example, clothespins for the preparation of string instruments. Out of this idea—the splitting of pure pitches into sound complexes—came the rather large *splitting* series of pieces for solo instruments. The title arises from the fact that the originally produced pitch is used as a trigger to initiate a second physical process. Because we still connect the result of the noise or the multiphonic sounds to the original pitch as the source I find it useful to call this process *pitch splitting*.

Applications such as plastic cups with or without marbles produce a more complex system of vibration. They have an extra resonant space and specific vibration characteristics. They serve more as extensions of the instrument than a simple preparation.

Finally, I use something that I call *new activators* to initiate the physical sound process. These are not preparations as I see them. Preparations change the sound when the instrument is used in the conventional way, for example by depressing a key on a piano. However, if the piano string is not set into motion by the hammer but is being stimulated by devices such as motors we are dealing with a whole new quality.

SB: The *new activators* sound a bit like the ever-present cappuccino creamers that were first used in improvised music and now seem to be ubiquitous when it comes to inside-piano playing. Does this device also fall in that category?

MM: Yes, and so does the EBow and so do the sonic motors that I use. Let me explain the use of this last device in detail using *splitting 36.1* as an example, which is a piano piece from 2011 and the first of four piano pieces that all explore different possibilities for defining the piano as a noise instrument. Why the piano? Because the piano is the paradigm of a pitch instrument: it is endowed with the richest possibilities of vertical and horizontal pitch combinations. And this is exactly the reason why it was the instrument composers worked with for so many centuries.

First I use two applications, each vibration system with its own physical characteristics: a plastic cup with glass marbles placed on the middle D and E♭ strings and a plastic half-sphere with marbles placed on the lowest C and C# strings (Figure 3, see Appendix, p.230).

The piece begins with the combination of an application and a motor: the plastic cup with marbles is set into vibration by a sonic motor.[3] After two measures this sound is combined with another noise structure when the piano hammers hit the strings on which the plastic cup with the marbles has been placed. The effect of this is that the vibrating strings set the plastic cup into vibration, but the glass balls irritate this vibration and a high noisy sound emerges. This mechanical distortion is amplified by the resonant space of the plastic cup. The result of all this is a sound that contains many elements, something which in my terminology I call a *sound complex*. For example, the *sound complex* mentioned above has four different characteristics (see Figure 4):

1. a continuous sound field resulting from the sonic motor together with the plastic cup and marbles. This sound field always changes slightly because the motor constantly moves on the string. It also includes the pitched sound elements of the motor itself;
2. the sounds of the different piano strings and their respective pitches, which are activated by the slight vibrations of the sonic motor;
3. the combination of the pitches of the keys in measure 4 together with the triplet pulse and dynamic changes, which are depicted by the volume graph.[4] This graph also controls the amount of noise; and
4. a sound layer that is introduced by the single impulses of the piano hammers in measure 6 and which are, in turn, distorted and amplified by the application of the plastic half-sphere.

---

3 A sonic motor is a motor that can be found, e.g., in ultrasonic toothbrushes; a Swiss-made model has proven to be the most musically enriching.

4 "*Dynamikkurve*" indicated in grey above the pitch staves (editor's note).

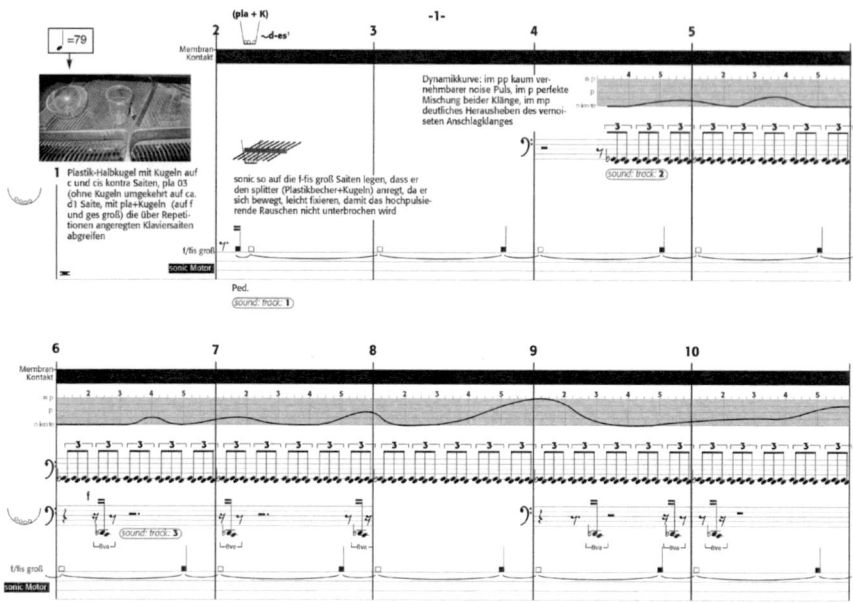

Figure 4 *splitting 36.1*, first page

SB: This is certainly fascinating and leaves one wondering how you find all these combinations. When you start a new piece, is there a systematic process you use in your approach to the instrument?

MM: When I explore the sounds, the fascination of the sounding material is mostly the starting point for a new piece. I make lists of possibilities, but not like a scientist who wants to explore all aspects. As early as the process of material exploration, I am already trying to work like an artist, with special focuses and interests, and always searching for the piece that will arise from this research.

Noise music or music generated with sound complexes has different production procedures to pitch-organized music. Because of the complexities in the sound itself, its construction takes place more within the sound than in pitched music, where you have known material and the composition process is all about the combination of the pitches, which can be planned beforehand. In non-pitch-organized music, analyzing material and formal processes through listening is probably more important than in pitched music.

From the beginning of my compositional work I was interested in exploring sounds and physical processes. Inspired by my studies of art history, it was clear to me that artists have to develop their own material, for example Renaissance painters who worked on new techniques of mixing and fixing color pigments. Personally, I was somehow not satisfied that most composers would adopt pitch material, and that their writing is then just a question of re-arranging the pitches.

SB: Does this mean that you were right from the start looking for another way to produce music? In the sense that you took a step back, looked at the production of contemporary classical music, and decided that there must be something else other than pitched music?

MM: Yes. From the beginning I wanted to produce non-pitch-organized music. It started with me preparing the cello, an instrument I studied a little bit at university, and so I could try out things at home. Suddenly a whole cosmos of sounds emerged from that extended instrument, which was propulsive and productive.

SB: So when you started to write music and you decided to write non-pitched music, did you not run into the problem of how to organize time? How do you organize time in non-pitch-organized music?

MM: More than organizing time, the problem is to organize the sounds: to create composed audible spaces that make sense, which are not arbitrary. However, this is always the point: to make music under our "everything is possible" conditions of today, to reach this point of "non-arbitrariness" [*Nichtbeliebigkeit*].

I listen to the sounds over and over again to find out what makes sense in an audible way; however, this is always connected with a few strong premises, for example that sound and silence have the same value or that sound complexes need no external relationship to be interesting in a musical sense.

SB: If I understand you correctly, you position the sounds by listening to them and finding the right place for them in a certain period of time and not by "fulfilling" a plan that you work out beforehand. Would you think that this is an approach called for by noise music, or that is even unique to noise music? Or to ask the question the other way around: is there a "traditional" composer whom you feel close to with this working method?

MM: With regard to my working method I feel closer to visual artists, some producers of electronic music, and improvisers than composers because I produce every sound I write by myself, which is normally not the working procedure of a composer.

SB: Was there a certain moment when it became clear to you that "this is it"? A eureka moment, so to speak?

MM: It has been a eureka experience over several years, actually.

SB: When we speak of noises in contemporary classical music, Helmut Lachenmann and his *musique concrète instrumentale* comes to mind. Here the keywords are the "emancipation of noise." In what relationship to this tradition do you see yourself?

MM: Composers in all periods have to find their position in relation to music history, society, and actual reality. My own concept of sound complexes is more an internal perspective on sounds, compared to Lachenmann's *musique concrète instrumentale*. My sound complexes are not a negation of anything; this is probably the most important aspect in my opposition to Lachenmann.

His turn to pitches in the late 1980s does not mean his concept of *musique concrète instrumentale* has failed, but that he himself could not go on with this kind of music anymore, which I clearly can understand. However, today we have different reasons for writing or producing non-pitched music, which has more to do with our common reality. Today, to negate tradition is no longer a political act.

Let's face it: the relation between art and politics is a difficult one. For example, the claim that all good art is inherently political waters down the concept of the political so much that it becomes diffuse. I am quite skeptical when art tries to be political in a straight sense and I am not sure we can change society directly through the means of an audible event. It is already a big claim to say that art can "regenerate the perception of reality," as Gabriel Orozco says. For me, regenerating or even changing perceptions of reality is a criterion of contemporary art, and art fails easily when it tries to be explicitly political or influential.

SB: Nonetheless, people might easily think of your music as political because of the tools that you use, for example by using a cheap plastic cup on very expensive instruments like pianos and violins.

MM: I use plastic cups solely as a means to produce the sounds that interest me. As I said above, I would not dare to use the word *political*. I would not deny, however, that art may have a *transformational energy* in the long term.

Nevertheless, this is a broad field and maybe we should stick more to the subject of noise music, which sometimes is regarded as anti-pitch, or pitch as anti-noise by those who like the opposition. Still, noise can also be regarded as a pitch category, because, strictly speaking, noise is just a combination of many pitches; the strict opposition between noise and pitch makes no sense. It shows, however, that the two parts of the spectrum can be compared to black and white in the field of colors: black is the absence of color and white consists of all colors mixed together in the same amounts. This has always been a model for the acoustic understanding of noise; however, it's too narrow. What we are talking about is not just the pure, homogeneous white noise that is normally used as paradigm for noise. Noise, or more specifically *sound complexes*, can have strong internal structures, and they can be the working field of a composer.

SB: Michael, you produce every sound you write yourself and you are probably the best player of those sounds as well. For example, in a recent concert by stock11[5] at the Acht Brücken Festival in Cologne, a new piece of yours was premiered: *specific objects, 11 min.* for four snare drums. Because the techniques you use are so unique—in this case the snare drums are played with sonic motors—it did not matter too much that not all the players were percussionists, as they could not fall back on learned techniques anyway. In fact, I was one of the players myself and I am a very bad drummer. Two questions sprang to mind: as you play your instruments best, why do you rely on other performers to play your music? And the second question: are those performers, who might for example be seasoned percussionists but who now have to learn a completely new and also very delicate instrument, not reduced to amateurs?

MM: Although I can play the sounds I cannot play them well enough to perform them. My music needs professional performers who know how to rehearse and to perform. I am not professional in this sense. I also need to write for larger groups and not just solo pieces. Finally, I like to collaborate with interpreters and let the sounds and the music go through another transmitter, so to speak.

To your second question: musicians who are playing extended instruments they have not studied—snare drums with motors in your example—are not amateurs. Only if I had asked you to play the snare with sticks would you

---

5 Musicians' collective, website accessed 16 June 2013, http://www.stock11.de/

have been an amateur, because there are clear standards for percussion players of how to do this professionally.

With regard to the new method of sound production in my pieces, what is needed is to develop an experimental approach and seven abilities: figuring out new playing techniques, aesthetic understanding, rhythmic precision, rehearsal experience, stage presence, transfer of body control, and the ability to have the perspective of a consistent musical interpretation. This is what separates a professional from an amateur in my view.

SB: Alright, that is a very clear statement. I would like now to ask you something I often get asked after I have played a piece of yours about the freedom of the interpreter. Earlier you have mentioned interpreters as transmitters. However, you have such a concrete vision of the sounds that you not only describe how they are to be produced but you also record them, and these recordings are an integral part of the scores. Together with instructions in the score that regulate the movement of the players in the pauses, it would seem like there is no freedom for the performers and that they are rather *executors* of a score than musicians who interpret a score. So, why do you prefer live musicians to recordings of your pieces?

MM: First of all, I definitely prefer a live performance because if you are working with the quality of sounds then the vibrating membrane of your stereo is completely incomparable to a sound emerging from a vibrating physical body.

Second, precision in the score does not necessarily limit the freedom of interpretation. If the score is a clear reference for the interpreters of what the composer is asking them to do, then they do not have to invest time in figuring out what the composer's ideas were. This gives the interpreters more time to find their *own* approach. And every interpreter has to take this freedom of their own approach. I actually expect this from a professional musician and it is in no way different from traditional music where pitch, rhythm, and dynamics are fixed too. I would also expect them to know about the history of interpretation or stylistic fashions.

In contemporary music there is usually not enough time to work on an attitude to the musical propositions of the composer. However, the fact that there is an unconscious interpretation of contemporary music must not be underestimated, simply because the composers and their interpreters live at the same time and share lots of acoustical and aesthetic experiences. There is an immediate understanding when it comes to the execution of the score, an understanding that is already an interpretation. This becomes clear years or decades later when musicians try to find an approach to this music again, but without having these common experiences anymore.

SB: Could you describe what changes when different performers play the same work?

MM: Every performer has a different set of skills, and the differences in the combination of the seven abilities that I mentioned above mark the difference between different performers.

SB: For the last question I would like to turn away from the practicalities of performing to those of music theory: how do you envision that your music, or noise music in general, can be analyzed?

MM: As I said before, non-pitch-organized music has a different kind of material: the material is the sound itself and its internal procedures. These are already musical statements that do not necessarily need an external relationship as in pitch-organized music.

As most theories—and they form the basis of music analysis—are pitch theories, new procedures are needed for the adequate analysis of a piece of non-pitched music.

For example, in order to notate sound complexes, composers need to write "multimedia" scores (print, audio, video, computer scores). With the aid of such a score an analysis could start from a detailed description of the sound complexes.

Describing the manufacture of the sound complexes might possibly be another analytical approach, combined with a description focusing on the various sound layers, the sound qualities, and the sound structures in order to define them *within* the sound. The description earlier of the sound structures in *splitting 36.1* is actually an example of such an approach. To me, this is an important strategy because it enables us to speak in detail about these sounds, which would otherwise remain opaque and abstract. In other words, separation and description are fundamental.

To put it simply, the first interest should be the internal relations of the material, and from there starts the analysis of the external relations, which would be an analysis including description and evaluation of how these sound complexes are composed. By this I mean, for example, the combinations, the sequences, the layers, the time extensions, and eventually the course or the process of the material. From there one can go on to the ideas of the piece as a piece of music.

However, many of these analyses will have to be done using different approaches before it is possible to find suitable methods. Developing a theory of the music of sound complexes is only possible through practice, and I assume that we will not end up with one general theory of noise music but with a completely new model of music theory.

NOISE IN AND AS MUSIC

Kasper Toeplitz

What is noise (music) to you? Why do you make it?

Noise music (or noise) is exactly what, some 25 years ago, or even earlier, made me choose "contemporary" music when deciding to become a musician, or even what made me want to play music full time, to live in music. I of course was listening to all kinds of things: jazz, rock, some classical, some more "specific" things (such as the "Canterbury school," or the English free jazz around labels such as Ogun, or musicians such as Chris McGregor) … I had this kind of all-around interest. Good enough; I liked it. But when I first heard compositions by people such as Krzysztof Penderecki, György Ligeti, Giacinto Scelsi, and a little later Karlheinz Stockhausen and Iannis Xenakis (and many others to follow: French and Romanian spectral music, Claude Eloy, Friedrich Cerha, François-Bernard Mâche, Luigi Nono …), OK, that was it—freedom in music. Much more than in free jazz, if you ask me. Freedom of structure.

But then in the following years contemporary music grew very old, and what once was freedom became a sure, tried, and safe formula. I was into music, listening to other things, discovering electronics (and MaxMSP), starting to get a little fed up of classical musicians and strict and safe ways of dealing with orchestras, string quartets, and this (very bureaucratic, very *passé*) world.

Then into my life came Japanoise, and noise in general, and music was exciting again! The long developments (more architecture than "proper music"), the pure physicality of the sounds (and not their "function," as you have in rock, jazz, etc.), and the passion.

In a way I am a noise musician, sure. But I also am a "contemporary-classical" composer. I don't mean that I do those two separate things, but that what I do you can call what you want. Maybe I am composing noise.

So, noise music to me? Freedom of invention. And why do I make it? Well, the answer is obvious.

But it is also good to add that it is one of the purest expressions of beauty, this music. Or pure ugliness, maybe, which makes it beautiful.

Materiality and Agency in Improvisation: Andrea Neumann's "Inside Piano"

Matthias Haenisch[1]

Pushing the boundary of material and continually innovating instrumental techniques are part of the history of jazz, and can be found in an even more radical form in the improvising avant-gardes of the 1960s and 70s.[2] The branch of improvised music that broke away from the jazz tradition was influenced by, among other things, the aesthetics of John Cage, which inspired far-reaching experiments involving instruments and material.[3] In this way the emancipation of noise also left its mark on improvisation, opening up new paths for the exploration of material, and it has had a formidable and lasting impact on performance practice in the field. The extension of musical material accompanied a correspondingly wider understanding of what can be considered an instrument, because, as a consequence of the leveling of the hierarchy between noise and music, "the need for a line between what is and what is not a musical instrument was firmly broken."[4] The resulting proliferation of musical objects and practices runs the gamut from extended instrumental techniques on traditional instruments, to the modification or dismantling of instruments, or even the construction of newly invented ones, to the innovative (mis)use of analog or digital recording and playback

---

1 Translated by Carter Williams.

2 Cf. Peter Niklas Wilson, *Hear and Now. Gedanken zur improvisierten Musik* (Hofheim: Wolke, 1999), 74.

3 Nina Polaschegg, "Reflexive Improvisation? Fortsetzung, Reflexion, Korrektur der 'Moderne' in der jüngsten 'frei' improvisierten Musik," in *Orientierungen. Wege im Pluralismus der Gegenwartsmusik*, ed. Jörn Peter Hieckel (Mainz, London, New York: Schott, 2007), 224.

4 Andy Keep, "Instrumentalizing: Approaches to Improvising with Sounding Objects in Experimental Music," in *The Ashgate Research Companion to Experimental Music*, ed. James Saunders (Farnham: Ashgate, 2009), 116.

equipment and the appropriation of found objects and industrial artifacts as sound sources. Many of these practices are unorthodox, manipulative, or deconstructive, and carry out "creative abuse"[5] on the existing world of things. With respect to musical material, the instrument is no longer a ready-made, general-purpose object. On the contrary, the instrument only emerges through a process of "instrumentalizing"[6] that aims to unleash the hidden musical qualities of virtually every object. This necessitates a paradigm shift in our concept of an instrument from a predetermined object to a performative act:[7] a fork, a clothespin or an electric fan only become instruments once they are integrated into a musician's performance practice. As improvisers strive to expand their individual sound repertoire, their unique, often homemade, and reconfigurable instrumentarium grows. The knowledge of the practitioner, her appetite for exploration, her aesthetic stance, and sometimes the implicit rules of her performance practice all begin to materialize in the process of constructing such an assemblage.

Andrea Neumann's "Inside Piano" is a representative product of this development. The instrument's construction and configuration bring together a multitude of instrumentalizing practices: the deconstruction of the musical instrument, the partial integration of other musical instruments such as violin bows and guitar pickups, the musicalization of everyday objects such as potato mashers and shaving brushes, and the creative (mis)use of audio equipment, for example using a mixing board as a sound source (Figure 1).

In this chapter I explore questions regarding the specific function and role of the instrument and the objects in Andrea Neumann's work.[8] I will

---

5 Keep, "Instrumentalizing," 116.

6 Keep, "Instrumentalizing," 113.

7 Keep, "Instrumentalizing," 113.

8 My investigation is based on Neumann's published and unpublished texts, as well as interviews and video recordings made over the course of a research project into performance practice in improvised music that has been in progress since 2011.

Figure 1: The Inside Piano overview (photograph by Manuela Stark)

attempt to show that the artifacts of the practice are at the same time among its actors; that is, they are performative participants in the improvisation. The significance of this observation is that it prompts the integration of natural and technological objects into the analysis of performative processes. Along these lines I would like to position myself squarely among those who question the widely held view that instruments are simply a tool, prosthesis, or extension of the musician.[9] In order to be able to observe the active role that things play in improvisation practice, I refer throughout the text to

---

9 Cf. David Borgo and Jeff Kaiser, "Configurin(g) KaiBorg: Interactivity, ideology, and agency in electro-acoustic improvised music," *Beyond the Centres: Musical Avant-Gardes Since 1950: Proceedings*, ed. C. Tsougras, D. Stefanou, and K. Chardas (2010), last accessed August 2, 2013, http://btc.web.auth.gr/_assets/_papers/BORGO-KAISER.pdf; Franziska Schroeder, "The voice as transcursive inscriber: The relation of body and instrument understood through the workings of a machine," *Contemporary Music Review* 25/1–2 (2006): 131–38.

several assumptions from Actor–Network Theory (ANT), which has its origin in the 1980s and was principally developed by Bruno Latour, John Law, and Michel Callon.[10] A central conceit of this theory is to examine the way objects, things, and technologies participate in social processes. In order impartially to determine the social meaning and function of things, ANT takes the generalized symmetry of human and non-human actors as a starting point. Within the framework of this theory any entity that is able to affect some sort of change is termed an actor (or actant). These can include people, animals, material objects, technical artifacts, concepts, and discourses. These entities become actors when they link to other actors to form networks. The emergence of such heterogeneous networks is a transformational process, in which the properties and activities of all human and non-human actors are equally involved and at the same time changed. By joining a heterogeneous network, human and non-human actors modify one another reciprocally. User and artifacts do not remain as they were before; they become something else. Such a network can therefore itself be seen and described as a new collective actor—a hybrid actor.[11] Against this background I would like to try to expose various aspects of the collaboration between human and non-human in Neumann's performance practice.[12]

---

10 Bruno Latour, *Reassembling the Social: An Introduction to Actor-Network-Theory* (Oxford: Oxford University Press, 2005); John Law, "Notes on the Theory of the Actor-Network: Ordering, Strategy, and Heterogeneity," *Systems Practice* 4/5 (1992): 379–93; Michel Callon, "Some elements of a sociology of translation: domestication of the scallops and the fishermen of St Brieuc Bay," in *Power, action and belief: a new sociology of knowledge?*, ed. John Law (London: Routledge, 1986), 196–223.

11 Bruno Latour, "On Technical Mediation–Philosophy, Sociology, Genealogy," *Common Knowledge* 3/2 (1994): 29–64; cf. Michel Callon, "Society in the Making. The Study of Technology as a Tool for Sociological Analysis," in T*he Social Construction of Technological Systems: New Directions in the Sociology and History of Technology*, ed. Wiebe E. Bijker et al. (Cambridge: MIT Press, 1987), 93: "[The actor-network] is simultaneously an actor whose activity is networking heterogeneous elements and a network that is able to redefine and transform what it is made of."

12 Cf. Borgo and Kaiser, "Configurin(g)," 4: "If we tentatively define electro-acoustic improvised music as real-time musicking involving humans, acoustic sound sources and spaces, and interfaces with electronics, then the practice appears to foreground (perhaps in somewhat equal measure) issues of human-human, human-machine and human-text (e.g., a computer program, or the 'media message' of performance) interactivity."

Figure 2: Andrea Neumann, Inside Piano (Background: Burkhard Beins, percussion, objects; not pictured: Valerio Tricoli, reel-to-reel tape recorder). Concert at KuLe, Berlin-Mitte, March 26, 2013 (still from video by Norbert Artner)

As was the case with many improvisers, Cage's work with the prepared piano was an important point of departure for Neumann in the search for new sounds "that convey a shorter history than that of a sound made by a pressed key."[13] However, in marked contrast to the now widespread practice of preparation and playing inside the piano, Neumann's instrument is one-of-a-kind. In its current form, it is made up of three playing areas: the "actual" Inside Piano consisting of an aluminum replica of a piano harp, an aluminum shelf for objects and preparations, and a mixer that is used for sound processing, mixing, and as a sound source. Neumann's recombination and continual further development of playing techniques and instrumental concepts from 20th-century and contemporary music history can be clearly

---

13 Andrea Neumann, "Playing Inside Piano," in *Echtzeitmusik: Self-defining a scene*, ed. Burkhard Beins et al. (Hofheim: Wolke, 2011), 203.

seen. These include exploiting and preparing the inside of the piano, techniques that first appeared in new music (Henry Cowell, John Cage) and then later in the collective improvisation of the 1960s (Nuova Consonanza); the use of everyday objects to generate sounds (Cage); idiosyncratic combinations of objects such as steel coils or electric fans with guitar pickups (Keith Rowe); and using the so-called no-input-mixer as a sound source (Toshinori Nakamura), a use which Richards has characterized as "bastardization."[14] In addition there are a number of original playing techniques that have originated on this instrument. Neumann began the development of the Inside Piano around 1995 at the same time that she started to emerge as a fixture in Berlin's *Echtzeitmusik* ("real-time music") scene. The instrument has not only been closely linked to Neumann's development as a professional improviser from the very beginning, but is also interwoven into the sociological fabric and aesthetic disposition of a scene of improvised and experimental music that came together in Berlin after the fall of the Berlin Wall. The development of the Inside Piano did not follow any *a priori* plan.[15] Rather, the instrument arose out of years of negotiation, communication, and interaction, which in the context of ANT can be termed a "translation," a process of building a network in the course of which the identities, capabilities, roles, and functions of human and non-human actors are redefined, displaced, and transformed in order to align them to a collective program of action.[16]

Initially it was economical limitations and their technical solutions that led to significant aesthetic developments.[17] Most of the venues in the free concert scene in Berlin at the time were underground, makeshift, and

---

14 John Richards, "32kg: Performance Systems for a Post-Digital Age," *Proceedings of the New Interfaces for Musical Expression Conference*, Paris (2006): 284.

15 Andrea Neumann, "Development of the Inside Piano" (Berlin: unpublished, 2008), 1.

16 Callon, "Sociology of translation," 196.

17 Neumann, "Development," 1.

repurposed spaces that as a rule did not have a piano.[18] Therefore Neumann had to come up with a transportable instrument just to be able to perform. Inspired by Zeena Parkins' prepared harp—both with regard to the lighter construction as well as the sonic possibilities—Neumann decided to do without the keyboard and the body of the piano and to work with just the naked plate, including strings and soundboard, from an original piano frame provided by the piano maker Bernd Bittman. But this now transportable instrument also brought certain limitations. Initially the cast iron frame was placed upright and leaned against a wall, making it extremely difficult to attach preparations to the instrument. Neumann sat on the floor wearing wool socks so that she could damp the strings with her feet. Later the instrument was placed on wooden blocks, and because it was not loud enough for certain performance situations, it was fitted with a guitar amplifier. In 2000, following Neumann's request, Bittmann, who had by that time already procured two original piano frames for Neumann, built a custom replica of a piano harp made of aluminum that was much lighter. This new instrument, which was designed above all to make transport easier, was also fitted with a pedal mechanism, which had been missing up until this point, as well as a shelf to hold preparations. However, due to the lighter construction the sound of the instrument was much weaker, which necessitated the installation of an optimized amplification system with an array of piezoelectric contact microphones and a mixer. This meant that the instrument's timbre became further removed from the "original" Inside Piano. On the other hand, the use of contact microphones led to intensified experimentation with subtle sounds and noises under the magnifying glass of extreme amplification.

Likewise, the mixer could be used as an additional sound generator and allowed for the simultaneous processing of parallel layers of sound as well as blending acoustic and electronic elements. Thus the continuous and often

---

18 Dietrich Eichmann, "Orte, Musiker, Ästhetiken. Die Berliner Szene," *positionen. Texte zur aktuellen Musik* 62 (2005): 22.

Figure 3: Contact microphone, knife, guitar pickup (photo by Norbert Artner)

pragmatic reconfiguration of the instrument paved the way for new and surprising possibilities to shape sound, with a marked effect on Neumann's musical bearing. The contact with an ever expanding number of objects and preparations gave rise to an inner soundworld connected to specific materials. And with the increasing awareness of materials, musical concepts, and new artistic goals began to crystallize, which Neumann then in turn transferred to further reconfigurations, new objects, and different preparations. Performer and artifacts found themselves entwined in a reciprocal process of perpetual transformation.[19] Through this process—over years of translation between the demands of performance practice, monetary constraints, craftsmanship,

---

19 Similar to Pickering's description of scientific instrument development, the development of the Inside Piano reveals itself as a *dance of agency*, "a performative, transformative and productive back and forth between human and non-human agency," see Andrew Pickering, "Ontological Politics: Realism and Agency in Science, Technology and Art," *Insights* 4/9 (2011): 3.

technological adaptations, instrumental reconfigurations, experimental preparations, aesthetic decisions and artistic impulses in a network of collaborating agents—work with amplified small sounds, probing the depths of the noise worlds in the gray area between acoustic and electronic, grew to become a characteristic feature of Neumann's personal style. The genesis of such a personal style would remain inscrutable against the backdrop of a dualistic view between an autonomous and intentionally acting subject and a determined object. Instead, performer and instrumentarium have become what they are through a "mutually constitutive process through which users, technologies, and environments are dynamically engaged in refashioning one another in a feedback loop."[20] In the course of this mutual configuration of performer and instrument, a hybrid actor has emerged whose aesthetic agenda is no longer reducible to the individual programs of action of the human and non-human actors who make up the network.

In the following account, Neumann gives a detailed insight into the relationship between performer and artifact, in which it becomes apparent how the materiality of objects, the construction of the instrument, physical abilities, practical knowledge, and musical imagination are equally involved in shaping the soundworld and developing a specific instrumental technique:

> A small bamboo rod (for the stabilization of plants), for example, has a particular radius. One must ascertain which strings the stick can be held between with relative firmness so that it stays put when rubbed. This procedure (with undampened strings and pine resin on one's fingers) leads to a voice-like sound. If there are places on the rod where the bamboo's skin is loose, then one must find out at which tempo and with which pressure this place should be brushed so that it makes a good sound. The

---

20 Borgo and Kaiser, "Configurin(g)," 2.

NOISE IN AND AS MUSIC

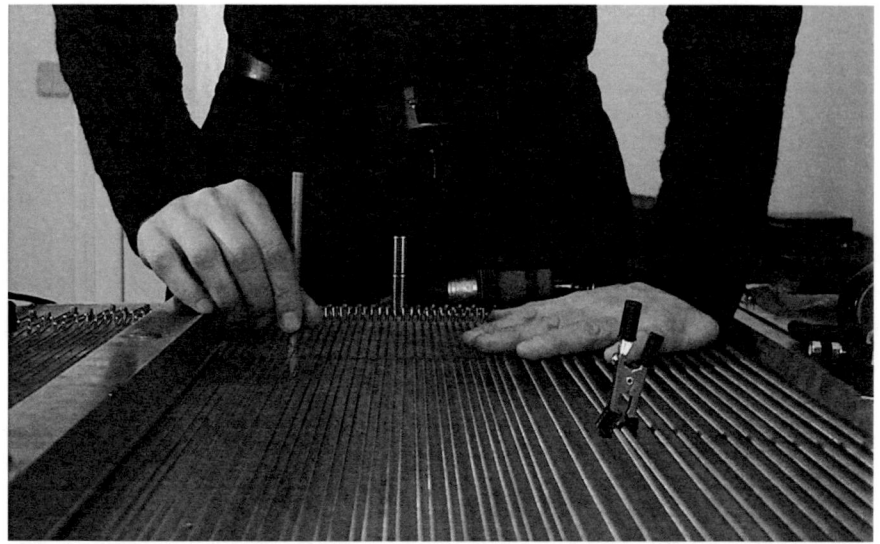

Figure 4: Neumann playing the bamboo rod (still from video by Matthias Haenisch)

damaged place can end up becoming the special quality of the preparation, since, for example, the air underneath the loose bamboo skin has an effect on the sound's pitch.[21]

Here we can clearly reconstruct the process of transformation and translation of disparate trajectories of individual entities into the program of action of the performer–instrument network. In an initially experimental setting, a bamboo rod is wedged between the strings in a manner similar to Cage's use of screws and nails. An attempt is made to coax a sound out of the bamboo rod, contrary to its purpose as a gardening tool, by rubbing it with a finger. Another artifact, violin rosin, also contrary to its intended use, is incorporated into the network so that the bamboo can be brought into a greater resonance through a stick–slip effect. The artifacts involved enable or hinder certain actions, demanding physical and technological adaptation

---

21 Neumann, "Playing Inside Piano," 207.

to their material nature and resistance. Slight variations in the position of the bamboo and differences in finger pressure and moisture lead to changes in the sound.[22] The materiality of body and artifacts enter into reciprocal coupling. At the end of this chain of operations a damaged spot on the bamboo rod serendipitously appears as an important factor in creating a truly original sound neither intended nor foreseen. Sound and playing technique cannot be traced back to an antecedent intention; rather they are a result of a contingent transformation of physical abilities, botanical secretions, historical references, horticultural intentions, mental states, material damage, and so on into an ephemeral network unified by a new goal. This process of interaction described by Neumann demonstrates the performative genesis of a characteristic sound and a particular playing technique as an emergent effect of a heterogeneous association between human and non-human actors.

Building on the work of Schroeder and Rebelo on the performer–instrument relationship, I understand this process of performative instrumentalization not to be a form of control but instead as creating an open and participatory environment[23] "in which the instrument or the device itself suggest specific ideas of their textures and in which the human body becomes acquainted with the 'thing' at hand by being able to test boundaries, negotiate subtleties and uncover certain threshold conditions."[24] This gives rise to a relationship between performer and object that can be described more specifically with the help of one of the central concepts of ANT, affordance.[25] Affordances

---

22 Neumann, "Playing Inside Piano," 207.

23 Cf. Pedro Rebelo, "Haptic sensation and instrumental transgression," *Contemporary Music Review* 25/1–2 (2006): 28.

24 Franziska Schroeder, "Caressing the Skin: Mobile Devices and Bodily Engagement," *Proceedings of the 5th International Mobile Music Workshop Vienna 2008*, accessed August 1, 2013, http://www.mobilemusicworkshop.org.

25 Bruno Latour, "Morality and Technology: The End of the Means," *Theory, Culture & Society* 19 (2002): 247–60; James J. Gibson, *The Ecological Approach to Visual Perception* (London: Lawrence Erlbaum Associates, 1986); Donald A. Norman, *The Design of Everyday Things* (New York: Basic Books, 1988).

are the particular performance characteristics offered by an object or an environment that permit or suggest certain uses while eschewing and hindering others. The twofold meaning of this concept is crucial. On the one hand there are invariant material and physical possibilities, obstacles, and efficacies that are inherent to the object of the bamboo rod. On the other hand possibilities for using the object only exist with respect to trained physical and mental skills, schemes of perception, expectations, and attitudes of the subject. Affordances are relationships of mutual dependency that subvert a clear subject-object dichotomy because they only become real in and through interaction. They reveal the relationship between materiality and corporeality to be one of performative productivity.[26] However, things, artifacts, and even natural objects are by no means passive matter that is only defined and shaped by practice. On the contrary the materiality of things, as an "agent of transformative efficacy,"[27] is always involved in creating and executing a practice.[28]

The concept of affordance refers not only to the transformative efficacy of the artifact but also the attitude with which Neumann confronts her objects. It also alludes to the meaning of the aesthetic criteria according to which Neumann chooses certain objects and preparations while rejecting others. As became clear in an interview I conducted with Neumann, she was guided in her research into the Inside Piano by criteria she developed in ongoing

---

26 R. Schmidt, *Soziologie der Praktiken. Konzeptionelle Studien und empirische Analysen* (Frankfurt/Main: Suhrkamp, 2012), 67.

27 Ingo Schulz-Schaeffer, "Technik in heterogenen Assoziationen. Vier Konzeptionen der gesellschaftluchen Wirksamkeit von Technik in Bruno Latours Werks," in *Bruno Latours Kollektive. Kontroversen zur Entgrenzung des Sozialen*, ed. Georg Kneer et al. (Frankfurt/Main: Suhrkamp, 2008), 110.

28 I am referring here to Barad's "posthumanist notion of performativity—one that incorporates important material and discursive, social and scientific, human and nonhuman, and natural and cultural factors." See Karen Barad, "Posthumanist Performativity: Toward an Understanding of How Matter Comes to Matter," *Signs: Journal of Women in Culture and Society* 28/3 (2003): 808. Cf. Mathias Maschat, "Performativität und zeitgenössische Improvisation," *kunsttexte.de/auditive_perspektiven* 2 (2012), 10, www.kunsttexte.de. While contemplating the performative genesis of musical material, Maschat seems to miss or deny the performativitiy and agency of the material.

cooperation with other musicians as one of the central figures of the Berlin scene.[29] This means that in addition to individual preferences, the aesthetic disposition—what Bourdieu would call the artistic *habitus*—of an entire community of practice was caught up in the construction of the instrument. Since the mid-1990s, Neumann has belonged to a circle of musicians within the *Echtzeitmusik* scene who saw the energy, gesture, and expressivity of contemporary free improvisation as becoming exhausted. The search for alternative concepts led to an investigation of the soundworlds of new music (e.g., John Cage, Morton Feldman, Giacinto Scelsi), the reduced performances styles of improvisation collectives of the 1960s (e.g., AMM, Gruppo di Improvvisazione Nuova Consonanza), and the post-digital aesthetics of contemporary electronica and electronic music. Toward the end of the 1990s, these influences coalesced into the so-called "Berlin Reductionism."

[This] manifested itself in a rather strict material selection, concentration on only some acoustic elements by eliminating the other, slowing down, reducing density of musical events, avoiding immediate reactions while improvising in a group, re-evaluating the relation between sound and silence, reducing dynamics range, all in order to be able to achieve more control and more focus on a chosen element.[30]

Due to the rejection of structural and formal relationships and because of the isolation of musical events, the singular manifestation of a sound or noise along with its materiality and corporeality became the basis of a radical aesthetics of presence. In this milieu the distinct and individual sound repertoires of the improvisers followed a collectively developed program of

---

29 Interview with Andrea Neumann, April 5, 2013; cf. Andrea Neumann, "Influences."

30 Marta Blažanović, "Berlin Reductionism—An Extreme Approach to Improvisation Developed in the Berlin Echtzeitmusik-Scene," *Beyond the Centres: Musical Avant-Gardes Since 1950: Proceedings*, ed. C. Tsougras, D. Stefanou, and K. Chardas (2010), last accessed August 2, 2013, http://btc.web.auth.gr/_assets/_papers/BLAZANOVIC.pdf; cf. Polaschegg, "Reflexive Improvisation," 226.

materiality. The musicians "were building common sound territories with focus on relatively quiet noises."[31] They tried to find material that would be rather "non-expressive, non-organic, and non-human, more machine-like, objective, and noise-like" that would recall everyday sounds "like those of washing machines, toilets flushing, heating, ventilation, or construction work."[32] The aesthetic principles of immersion, "zooming into the sound,"[33] and "acoustic microscopy"[34] became the categories for the selection, exploration, and molding of sound-categories that were translated into the construction of the Inside Piano, its extensions, and its playing techniques.

The form of the instrument and its reduction to the inside of the piano already seem to be a material exemplification of the immersive focus characteristic of Neumann's performances. In a similar way this "broken" relic of an earlier tradition corresponds to an aesthetic of "damaged sound," which Gottstein has identified as a distinctive feature of the Berlin scene.[35] While hardly suitable for playing melodic and harmonic structures, Neumann's rudimentary piano has been optimized for the production of a wide range of precisely articulated noises and sound textures. The aesthetic principle of acoustic microscopy finds its physical manifestation in the close positioning of contact microphones at various points on the instrument and the invasive amplification of the quietest sounds on the Inside Piano: minimal movement, steel wool sliding over a guitar pickup, scraping sandpaper with jazz brushes (Figure 5), or letting a whisk vibrate while attached to the aluminum shelf

---

31 Blažanović, "Berlin Reductionism."

32 Andrea Neumann, "Statement," in *Reduktion: zur Aktualität einer musikalischen Strategie*, ed. Peter Niklas Wilson (Mainz: Schott, 2003), 129; cf. Marta Blažanović, "Berlin Reductionism," 4.

33 Gisela Nauck, "Im Klang arbeiten. Innovationen in der aktuellen Improvisationsszene," *kunsttexte.de/auditive_perspektiven 2* (2012), last accessed August 2, 2013, http://edoc.hu-berlin.de/kunsttexte/2012-2/nauck-gisela-3/PDF/nauck.pdf.

34 Björn Gottstein, "An Aesthetic of Refusal," in *Echtzeitmusik: Self-defining a scene*, ed. Burkhard Beins et al. (Hofheim: Wolke, 2011), 152.

35 Gottstein, "Aesthetic of Refusal," 152.

MATERIALITY AND AGENCY IN IMPROVISATION

Figure 5: Jazz brushes on sandpaper (still from video by Matthias Haenisch)

all yield extremely heightened and, at the same time, somehow alien "interior views" of the sound.

The extreme diversity, internal articulation, and nuanced dynamics of many sounds generated on the Inside Piano are most effective in musical contexts where high density and volume has been abandoned in favor of absolute transparency and silence, which allows the presence of the sound to take center stage. The "bruitist" material aesthetic is represented by the multitude of objects that produce automatic, repetitive, and "mechanical" structures: hand-held electric fans, pulsating feedbacks, vibrating magnets, or oscillating knives, forks, etc. (Figure 6).

Many of these sound and object combinations came out of connections to the sound repertoire of a fellow musician:

> The material that I work with can mostly be traced back to my experiences playing with other people. It is really the case that you are engaged in a

NOISE IN AND AS MUSIC

Figure 6: Forks (photo by Norbert Artner)

process of mutual influence. For example it often happens that in rehearsal or even while playing a gig you notice some particular sound and think that you would actually need a specific type of material in order to be able to react to that sound. And so you go looking for it.[36]

Certain sounds that found their way into Neumann's repertoire came out of her work with the ensemble Phosphor around 2000. A "fan-belt-like" noise that Neumann generates by bowing a clothespin attached to the aluminum shelf with a violin bow (Figure 7) reminds her of a signature sound of her fellow ensemble member, the tubist Robin Hayward. The various noisy sounds that Neumann works with—e.g., bowing the edge of the aluminum shelf with a violin bow—also represent a direct material connection, especially to

---

36 Interview with Andrea Neumann, January 21, 2013.

MATERIALITY AND AGENCY IN IMPROVISATION

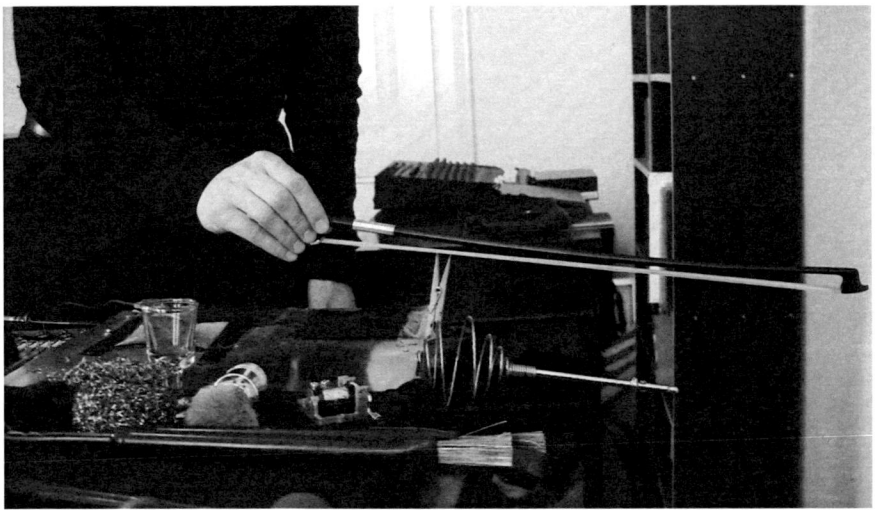

Figure 7: Bowing a clothespin attached to the aluminum shelf (still from video by Matthias Haenisch)

the breath sounds used by Hayward and the trumpeter Axel Dörner, or the static noise of the guitar amplifier employed by the guitarist and electronic musician Annette Krebs.

Neumann's instrumentarium is a materialization of her praxis: the continuous interaction with sounding objects and preparations, the emergence of playing techniques, and the exchange, appropriation, and transformation of musical material have embedded an aesthetic disposition and artistic preference in the technical configuration of the instrumentarium. This process is termed "inscription" in the context of ANT.[37] Such "scripts" fixed in artifacts are materializations of implicit knowledge that is drawn from the relevant contemporary practice. Likewise they can accommodate spaces for potential further use, implicit visions of future praxis. Through continual interaction and mutual transformation of body and artifacts, the

---

37 Madeleine Akrich, "The De-Scription of Technical Objects," in *Shaping Technology, Building Society: Studies in Sociotechnical Change*, ed. Wiebe E. Bijker and John Law (Cambridge: MIT Press, 1992), 205–24.

instrumentarium becomes a "materially stabilized choreography of body postures and movement."[38] The material inscription of the artifact takes on a central role in the performative (re)production of the performer and is an active part of her performance persona because "people are who they are because they are a patterned network of heterogeneous materials."[39]

Moreover, instruments—like all technology—are "society made durable."[40] They stabilize the aesthetic and social relationships between musicians in that they preserve collective memories of experiences and keep them available for upgrades and advancements. The materialization of praxis in an instrument in no way entails limiting the performer to strict norms or rules based on technical requirements, for example to an exclusively "reductive" attitude to playing. With this in mind ANT emphasizes the flexibility, fluidity, and permanent malleability of technology. As such, Neumann's obvious preference for separately excited or self-excited oscillating processes opens up the possibility of activating several objects simultaneously, which can lead to multilayered and polyphonic structures beyond "reductive" procedures. Recordings with Krebs from 2000 document the wide range of the instrument's potential and showcase Neumann's sounds in complex and lively interactions.[41] Both the "return to rhythmic and gestural patterns, ... comprehensible formal structures, linear progression"[42] that Polaschegg observes in Neumann's *modus operandi* as well as her growing interest in sometimes massive electroacoustic feedback drones are also effects of a continued investigation of material affordances beyond reductive strategies. Only Neumann's recent increasing

---

38 Schmidt, *Soziologie der Praktiken*, 154.

39 John Law, "Notes on the Theory of the Actor-Network: Ordering, Strategy and Heterogenity," *Systems Practice* 5/4 (1992): 383.

40 Bruno Latour, "Technology is society made durable," in *A Society of Monsters: Essays on Power, Technology and Domination*, ed. John Law (London, New York: Routledge, 1991), 103–31.

41 Annette Krebs and Andrea Neumann, *Rotophormen* (charzima 009, 2000).

42 Polaschegg, "Reflexive Improvisation," 229.

interest in the reintegration of tonal material really pushes the boundaries of the instrument and could, as she indicated in another interview, necessitate extensive modification or expansion of the Inside Piano in order to fit the instrument with a rudimentary keyboard.[43] Thus the performer–instrument network is not in a state of increasing fossilization but rather in a process of continuous performative reconfiguration.

The agency of artifacts is by no means simply restricted to the configuration of the instrumentarium and its playing techniques, but is also seen in the playing process of improvisation. The sounds that Neumann has worked out are explored to the finest detail of texture, and the behavior of objects and preparations are familiar from targeted practice and repeated performance. Having said that, not only the search for but also the reproduction of noises and sounds is influenced by many unpredictable factors. Occasionally one has to surrender oneself to an

> anarchic or chaotic process, whose results resist being fixed in any sort of notation: for example I take a small pickup and run it over a string. Which part of the pickup first comes into contact with the string? And when I move it, what part then in next touches the string? These small alterations generate a vast spectrum of pitches and sounds. The musical material is recalcitrant to reproduction.[44]

Often it is the material properties of an object (i.e., a steel wool ball, Figure 9) that are unsuitable for achieving a precise sound reproduction, or the fragility of a combination of objects (i.e., a vibrating bar magnet that is placed on a piano string or on a guitar pickup) that refuse to let themselves be subject to an exact way of playing.

---

43 Interview with Andrea Neumann, January 21, 2013.

44 Gisela Nauck, "Alte Fragen neu: Form und Inhalt. Ein Gespräch von Gisela Nauck mit Annette Krebs, Andrea Neumann, Serge Baghdassarians, Burkhard Beins und Axel Dörner," *positionen. Texte zur aktuellen Musik* 62 (2005): 13.

NOISE IN AND AS MUSIC

Figure 8: Scratching strings with a guitar pickup (still from video by Matthias Haenisch)

Figure 9: Steel wool ball (photo by Manuela Stark)

This resistance of the artifacts is a fundamental aspect of their efficacy as agents. In this context the instrument acts as a counterpart "that cannot be mastered, that has its own momentum" and that can behave out of line with expectations: "if I do something three times in a row, it sounds different every time. It is extremely difficult to predict a result and to create structures where I know exactly when something will happen."[45] From a socio-technological perspective, such resistance and indeterminacy are clear indications for the agency of artifacts, as an interactive communicative relationship between human and technology comes into being when an artifact behaves contrary to expectations, that is, when it acts contingently.[46] It is in this sense that Latour designates actors

> as obstacles, scandals, as what suspends mastery, as what gets in the way of domination, as what interrupts the closure and the composition of the collective. To put it crudely, human and nonhuman actors appear first of all as troublemakers. The notion of *recalcitrance* offers the most appropriate approach to defining their action.[47]

However, the contingency and resistance of the material is not a matter of insufficient playing technique that could or should be completely mastered with practice and increased control. Rather, it plays a fundamental aesthetic role. Repetition creates deviations that can be taken as impulses and triggers for the performance. Thus the reproduction of sounds becomes a performative production of differences that are capable of decisively altering the course of an improvisation. Accordingly the material resistance and indeterminacy of

---

45 Interview with Andrea Neumann, January 21, 2013.

46 Werner Rammert, "Technik in Aktion: Verteiltes Handeln in soziotechnischen Konstellationen," in *Autonome Maschinen*, ed. Thomas Christaller and Josef Wehner (Wiesbaden: Westd. Verlag, 2003), 289–315.

47 Bruno Latour, *Politics of Nature* (Cambridge: Harvard University Press, 2004), 81.

the instrument demands a specific type of listening: "because the material is recalcitrant, I have to always stay curious. If it always happened as I expected it, I wouldn't need to listen so carefully anymore."[48] A consistently high level of attention and sharp motor reflexes in response to small and large deviations are essential so that Neumann can keep pace with the effects that these deviations can have on the course of the improvisation by being prepared to make manual adjustment at any moment.[49] The often mentioned appreciation that many improvisers have for productive disorientation, happy accidents, and unpredictability is inscribed into the materiality of Neumann's instrumentarium.[50] Because of their material stubbornness, the artifacts call for the presence of the performer, hinder routinization, and act as agents in a collective creativity.

> It is precisely the indeterminacies that lead me down unexpected twists and turns. It can be a pulsation in feedback that I hadn't intended, which I then might try to amplify or react to with another sound, and this then takes me down a new path. And sometimes the instrument does something that is not only unexpected but much better than what I actually wanted[51]

The repurposing of things and artifacts is an encounter with their alterity. It facilitates the productive inclusion of the impetuous resistance of the object. It recognizes in technology a force that transforms the human element and gives the improviser cause to ask, "are we performing the technology or is

---

48 Interview with Andrea Neumann, May 25, 2013.

49 Cf. Keep, "Instrumentalizing," 118: "During the process of instrumentalizing the performer needs a … type of listening that is responsive of real-time activity."

50 Annette Krebs et al., "Zum kreativen Potential des Fehlers: Statements von KomponistInnen und MusikerInnen," *positionen. Texte zur aktuellen Musik* 79 (2009): 32–39.

51 Interview with Andrea Neumann, May 25, 2013.

it performing us?"[52] Apparently the practice of instrumentalizing, which emerged in the course of the expansion of musical material, offers instruments and artifacts recognition as material actors in the improvisation. However, the point is not to equate human and non-human actors, nor to detract from human artistic achievement and responsibility, but rather to integrate the influential presence and meaning of artifacts and objects into our notion of artistic ingenuity. Above all it is a part of Neumann's artistic impetus (as well as that of many other musicians) to value and expose the potency and obstinacy of objects in a creative way. A consideration of the agency and performativity of material can already serve in the observation of the reciprocal coupling of human and non-human: in the praxeological classification of the artifact as a physical mediator for transporting implicit knowledge and a collective history of interaction, and moreover in the identification of the program of action and aesthetic principles that are inscribed in the artifact.

By classifying instrumentaria and objects as agents of transformative efficacy and contingently acting participants, it becomes clear, however, that they can effectively influence the course of an improvisation. In this context artifacts act as delegates of a performative aesthetic, whose potential for innovation is not only derived from interaction among humans but also benefits from human–non-human interaction. For the analysis of performances of improvised music this should mean that, alongside corporeality, the materiality of things should also be recognized as an agent of performative practices, and the number of actors to expect and observe should be increased, as

> to distribute roles from the outset between the controllable and obedient object on the one hand and the free and rebellious human on the other is to preclude searching for the conditions under which ... one can ... make

---

52 John Robert Ferguson and Robert van Heumen, "Whistle Pig Saloon: Performing Technologies," *Leonardo Music Journal* 20 (2010): 12.

these entities exchange among themselves their formidable capacity to appear on the scene as full-fledged actors.[53]

Figure 10: Andrea Neumann, Inside Piano; Burkhard Beins, percussion, objects; Valerio Tricoli, reel-to-reel tape recorder. Concert at Ausland, Berlin-Prenzlauer Berg, May 15, 2013 (still from video by Matthias Haenisch and Ronny Zimmermann)

---

53 Latour, *Politics of Nature*, 81.

Franck Bedrossian

What is noise (music) to you?

"Noise" is the name that we give to a sound that does not have any *a priori* musical function and that is, for this reason, identified as non-musical. From that point of view only, speaking of noise music would be nonsense, as each sound perceived as musical is not a noise anymore.

But if one tries to define the main characteristics of this category of sounds, then two traits clearly emerge:

1. Noises are considered non-musical because their acoustic structure does not refer to—even contradicts—the usual hierarchy between musical parameters in the Western musical tradition, in other words all sounds that do not project pitch as a prominent quality.
2. Sounds that are considered noises are actually produced by sources and devices (nature, computers, urban life, etc.) that are not supposed to integrate into musical discourse because their anecdotal aspect—or cultural charge (sounds of everyday life, white noises of all kinds, etc.)—is immediately associated with non-musical situations in the mind of a listener.

Then one could say "noise music" is a music in which sonic material is made of complex sounds that deny pitch to be the most prominent dimension, and that despite an extra- or non-musical connotation are given structural functions within a network of musical relationships.

Why do you make it?

I do not consider myself a composer of "noise music." However, integrating the acoustic and aesthetic potentialities of all kinds of acoustically complex sounds in my music does represent one of the main stakes of my research.

Knowing that the instrumental world, as it has been conceived within the Western tradition, tends to exclude or hide complex sounds from instrumental possibilities, an approach that includes them and gives them musical functions allows composers to transcend categories, to play with the thresholds of perception, and to modify or even subvert the hierarchies within musical discourse.

Eventually, the integration of different territories of what we call "noise" represents a compositional and aesthetic challenge that might help to create different types of musical pleasure and emotions.

Noise-Interstate(s): toward a subtextual formalization

Joan Arnau Pàmies

Introduction: noise as transcender

Humans perceive noise as an indiscernible entity. Noise is problematic due to its intrinsically chaotic information, yet paradoxically fascinating because it presents great levels of richness and complexity. We view it pejoratively when it interferes with our logic and predictions, but at the same time we recognize its ability to avoid simple patterns and to embrace unpredictability. From a human perspective, reality is an extremely intricate network of relationships that we acknowledge, disregard, presage, judge ... the list could go on. Yet noise acts upon this reality as a chance operator: it transcends control, hierarchy, and imposed power.

For the past few years, my works have mostly been concerned with the attainment of formal richness. On a surface level, these pieces present highly active sonic materials that span short periods of time before they undergo multiple transformations. This is the result of employing multi-layered compositional processes that attempt to decouple instrumental means of sound production. I have found that this aesthetic tendency allows for the development of complex formal relationships on a deeper level of construction—this is an essential aspect of my aesthetic interests. A major consequence of my personal approach to instrumental decoupling is the intrusion of timbral noise, which is mostly due to paradoxical physical actions applied to a number of different instrumental techniques. Although this chapter does not primarily focus on this type of timbral noise, it delves into the role of a parallel "noise" in my notational practice, particularly the space between notational data and the execution of that data. I have termed this space the *noise-interstate*.

Information Theory and the Noise-Interstate

Before I examine the peculiarities of the noise-interstate, I shall describe some key concepts that originally appeared within information theory, which will be discussed later in other contexts. Information theory gained substantial presence within the scientific community after the publication of Claude E. Shannon's 1948 paper "A Mathematical Theory of Communication." That paper was Shannon's attempt to describe the intricacies of communication systems and their basic elements (information source, message, transmitter, signal, noise source, received signal, receiver, and destination) and also offered mathematical explanations of concepts that would be highly influential in future research.[1] For the purposes of this essay, two of Shannon's original ideas will be examined: *noise* and *equivocation*. Shannon portrays noise as an element that influences negatively the proper reception of the message: "If the channel is noisy it is not in general possible to reconstruct the original message or the transmitted signal with *certainty* by any operation on the received signal E. There are, however, ways to transmit the information which are optimal in combating noise."[2] For Shannon, *noise* is simply an undesired phenomenon that needs to be terminated. As literary critic N. Katherine Hayles points out:

---

[1] In music, information theory has been significantly influential within the domain of theory, especially between the late 1950s and early 1970s. As Vanessa Hawes writes, "[t]he idea of information theoretical studies of music imagines that it will give people a greater insight into the workings of human communication systems. The use of information theory assumes that communication can be used as a metaphor for music. This metaphor was used to a lesser or greater extent of sophistication. Using it in a quantitative way, as most of these writers did, naturally evolved into the use of computers in music analysis." ("Number Fetishism: The history of the use of information theory as a tool for musical analysis," *Music's Intellectual History*, ed. Z. Blazekovic (New York: RILM, 2009), 836) Some of the writers referred to by Hawes are Joel E. Cohen, Edgar Coons and David Krachenbuchl, Lejaren Hiller and Calvert Bean Jr., and Joseph E. Youngblood. Most of their theoretical work used concepts from information theory in attempts to discuss music on a more objective level.

[2] Claude E. Shannon, "A Mathematical Theory of Communication," *Bell System Technical Journal* 27 (July and October, 1948): 407.

NOISE-INTERSTATE(S): TOWARD A SUBTEXTUAL FORMALIZATION

When Shannon gave information theory its definitive formulation in 1948, his main concern was how to get a message through a channel with the least possible distortion. Whenever a message is transmitted, some noise inevitably intrudes—snow on a television set, static on a radio, blurred type or misprints in a book. Shannon called this noise *equivocation* and defined it using the same mathematical terms he used for information. In fact the equivocation is information, but from the sender's point of view it interferes with the intended message. Shannon therefore gave the equivocation a negative sign in his equation, indicating that it should be subtracted from the received message in order to get the original information back again.[3]

Equivocation can also be defined as an unwanted surplus of information (which does not belong to the original message intended for transmission) or, in Shannon's own words, "the average ambiguity of the received signal."[4] Equivocation is noise.

It took almost 30 years for the nature of equivocation to be problematized. It was a French Algerian biophysicist, Henri Atlan, who envisioned another aspect to the equation. Equivocation, according to Atlan, does not necessarily have to be the amount subtracted from the information that has been sent through a channel; it can also function as a beneficial addition to the whole system of communication—it all depends on a matter of perspective.[5] As Hayles points out:

> in some instances the equivocation's sign should be positive because it is possible to imagine a viewpoint in which the equivocation is constructive

---

3 N. Katherine Hayles, "Two Voices, One Channel: Equivocation in Michel Serres," *SubStance* 17, no. 3 (1988): 3. Emphasis added.

4 Shannon, "A Mathematical Theory of Communication," 407.

5 Henri Atlan, "On a Formal Definition of Organization," *Journal of Theoretical Biology* 45 (1974): 295–304.

rather than intrusive, for example *when it causes a system to re-organize itself at a higher level of complexity.* This ambiguity in equivocation's sign is related to the observer's position within the channel. If the observer is at the source, she knows what the 'correct' message is and will consequently regard the equivocation as interference. By contrast, if she is the receiver, she can see the effect of the noise on the system and will therefore observe instances in which it has positive effects.[6]

Atlan opened the door toward an understanding of noise as a potentially beneficial byproduct of the act of communication: "[o]ne thus sees how a positive 'organizational' role for noise could be conceived within the frame of information theory without contradicting the theorem of the noisy channel."[7] Furthermore, by drawing a parallel between information theory and biophysics (his main field of research), Atlan suggests that "where the evolution of species is concerned, no mechanism is conceivable outside of those suggested by theories in which random events (chance mutations) are responsible for evolution toward greater complexity and greater richness of organization."[8] Thus, Atlan notes not only that unexpected events are responsible for the ability of a system to reorganize itself, but also that life itself has developed toward higher degrees of complexity after *chance* mutations—that is, randomness. Noise, inasmuch as it is a phenomenon of chance, seems to fit this idea very well. For this reason, the inclusion of noise in my work has become an inevitable goal.

---

6   Hayles, "Two Voices, One Channel," 3–4. Emphasis added.

7   Henri Atlan, "Noise as a Principle of Self-Organization," in *Henri Atlan: Selected Writings*, ed. Stefanos Geroulanos and Todd Meyers (New York: Fordham University Press, 2011), 104.

8   Atlan, "Noise as a Principle of Self-Organization," 110.

NOISE-INTERSTATE(S): TOWARD A SUBTEXTUAL FORMALIZATION

The Noise-Interstate as a Self-Generating Entity

The noise-interstate is a psychological state that exists within the performer's psyche during the interpretive process of my work. Its primary goal is to contribute to the elaboration of multiple potential sonic outcomes whose particularities share certain essential characteristics among themselves and in relation to the original musical score. While the identity of the resulting music stays intact, the noise-interstate diversifies the potential interpretations of the work, thus presenting a greater degree of sonic variation across a number of performances. Therefore, its influence during the interpretative process of the score can only be perceived after listening to multiple performances or versions of the same piece. What I propose is an approach to notation that allows the noise-interstate to intervene.

Deliberate equivocation attempts to reconsider the original notion of certain material and formal constructions in such a way that the process of interpreting the score triggers a procedure of reorganization that is partly unintended during the early stages of the work's formalization. Furthermore, deliberate equivocation inherently contributes to the understanding that composition and performance are categorically different activities; the former is principally concerned with both the creation and organization of notational signifiers, while the latter is intrinsically closer to interpretational operations in relation to the score. The type of reorganization that takes place during performance is not left completely to the performer's discretion but instead assists in the redistribution of potential sonic relationships in such a way that the piece is dimensionalized but its integrity remains intact. The performer is thus capable of creating degrees of variance, which may suggest unaccountable formal paths that transcend both the peculiarities and the original implications of the compositional process of the piece. This is, however, an issue that often needs to be reconciled—the level at which the noise-interstate (triggered by deliberate equivocation) becomes an undesired influence can be problematic. Nonetheless, a competent use of deliberate equivocation on behalf of the performer might produce a set of two parallelogramic windows: a quantitative one, which sparks a field of intermediate states between slightly and highly unpredictable results, and a qualitative one, which provides the possibility of unforeseen formal trajectories. The noise-interstate is thus a truly experimental mechanism, for it attempts to raise questions that would otherwise not have been formulated.

Deliberate Equivocation: a methodology

I must now briefly explain the nature of my recent notation before delving into the specifics of deliberate equivocation. On a basic level, this notation is divided into three fundamental domains: *projection surfaces*, *parameterized objects*, and *temporal organization*. One needs to imagine that these are separate layers (each with its own attached set of processes) that are superimposed onto each other. They contribute to an array of multi-level paradoxes that will be discussed below.

The *projection surface* (Figure 1) functions as a delimiter of vertical space. In other words, it provides a specific area where other materials acquire a defined contextual situation. The *parameterized objects* (Figure 2, see Appendix, p.231) consist of a number of signifiers that convey certain physical motions that the performer applies to her or his instrument. For instance, they indicate

Figure 1: Projection surface in *[k(d_b)s]*, p. 7

technicalities in relation to instrumental means of sound production such as amount of bow pressure, position of the left hand on the fingerboard, and bow speed. *Temporal organization* (Figure 3, see Appendix, p.231) provides an insight into the specific nature of the temporal speed and division that has been developed throughout the process of composition. Once these three domains are juxtaposed onto each other, one can see the final disposition of the score (Figure 4, see Appendix, p.232).

In recent pieces, I have developed two compositional techniques that contribute to the process of deliberate equivocation. These are *temporal displacement notation* (TDN) and *perspective notation* (PN).

TDN compresses and/or expands the horizontal scale to which the parameterized objects and the projection surfaces belong. In other words, its main purpose is to re-establish the relationship between the numerical information on the timeline and its visual counterpart in multiple states. TDN allows the performer to redefine the nature of her or his own comprehension of the notated timelines, thus sonically producing slower or faster surface *tempi* in relation to the translational possibilities of the notated materials.

[IVsax(op_VIvln/c)] is a piece for solo saxophone with optional parts for violin and/or cello. One can see from Figure 5 (see Appendix, p.232) that the saxophone part is visually the largest on the score, and also that it directly maps onto the two upper timelines (A and B). In fact, these timelines correlate only to the saxophone part. What this means is that the "outer" performers (violin and cello) do not translate the notational materials into sound based on their own interpretation of time but must rather gravitate toward the saxophone player's sense of time. The saxophone part functions as the temporal center of the work, whereas the violin and cello parts operate as satellites. This potentially increases the amount of entropy within the channel between score and performer. The main attribute that deliberately contributes to equivocation is the dislocational praxis in the relationship between the timelines and the spatial situation of the outer parts. For instance, the cello part shown in Figure 5 (see Appendix, p.232) is meant to be performed during

timeline B of the saxophone part⁹ and should span the same duration (i.e., 10 seconds), though this is visually contradictory since the area occupied by the cello part on the score is substantially smaller than that of the saxophone part. Thus, the cellist is left with no visual references, in that she or he cannot base her or his interpretational decisions on the part's vertical alignment with the saxophone part. Such a disassociation raises many interpretational questions that need to be solved during rehearsal.

Another type of TDN appears in *[k(d_b)s]*, for solo double bass. Note the development of the temporal organization of this example, as indicated by the green timeline. The areas of the double bass part that are left white (as opposed to those that present different shades of green) contain the original spatiotemporal scale on which the parameterized objects are superimposed. The horizontal length occupied by the representation of one real second within the white areas is different than the length that one second will take within a green area (Figure 7, see Appendix, p.233).

Indeed, each particular shade of green indicates a specific speed at which the parameterized objects pass.[10] One major consequence of this technique is closely related to a type of parameterized object in particular. The bottom area of the projection surface in Figure 7 displays information related to the range of the instrument.[11] Within that area, one can see several groups, each consisting of an array of vertical lines of different lengths. The bottom ends of these lines indicate the approximate placement of the finger onto the string indicated by the attached Roman numerals—that is, the lower the notated end is within the range, the lower the sound produced should be in relation to the fingered string. Although one could count the exact number of vertical lines

---

9   The color code provides this information. Notice that the boxed letter B has a thin green rectangle to its left. This is the same shade of green that applies to the top of the cello part. For this reason, the cellist is only supposed to perform during the saxophonist's timeline B (7:02–7:12).

10 The darker the shade, the faster the speed, and vice versa.

11 The top line represents the placement where the strings meet the end of the fingerboard (high), while the bottom line signifies "open string" (low).

NOISE-INTERSTATE(S): TOWARD A SUBTEXTUAL FORMALIZATION

per group, the truth is that each collection should simply be performed as an extremely fast accumulation of left-hand articulations. Since one second will have differently notated lengths according to its surroundings (e.g., whether its background is white or green), the performer should keep in mind that the exact number and quality of the articulations produced after interpreting the groups of vertical lines fall into a realm that exists beyond the original text. The disconnection between the score and the performer is highly evident, thus contributing to the emergence of deliberate equivocation.

In contrast, PN, explores the potentially unstable peculiarities available via the juxtaposition of the parameterized objects onto the projection surface. PN, however, does not directly transform the characteristics of the objects, but instead redefines the topology of the projection surface to a lesser or greater degree. Such redefinition of the delimiting space provided by the score drastically affects the signification of the parameterized object to the extent that its own interpretational denotation reshapes the performer's views of the notation. Therefore, PN increases the entropy within the channel between score and performer, allowing a completely new level of textual ambiguity (or *interstateness*) to flourish and pushing the performer to recreate mentally a new subtext based on the original score that has the potential to trigger relatively unexpected sonic results, hence the multi-dimensional identity of the musical work.

My piece *[IVflbclVIvln/c]*, which was my first attempt to develop PN, presents a quite rudimentary approach to this technique. Figure 8 (see Appendix, p.234) exposes how the boundaries of the projection surface are no longer horizontally equidistant but instead lead independently in different directions. For instance, the bottom area of the projection surface (whose parameterized objects indicate the vertical position of the fingers of the left hand on the fingerboard of the cello) contributes to the steady reduction of its own area, thus readjusting the contextual scale to which the parameterized objects belong.

Figure 9 provides a more specific insight into this phenomenon. A simple calculation of the starting distance between the long horizontal line in the

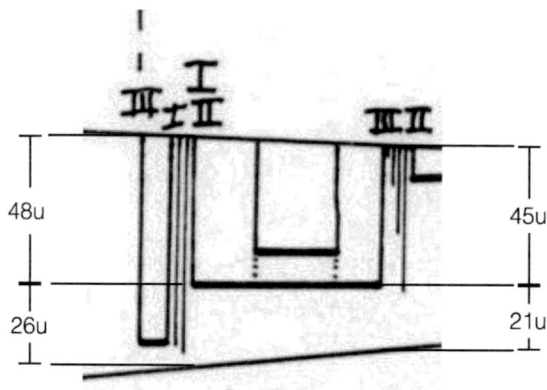

Figure 9: *[IVflbcIVIvIn/c]*, p. 1, cello part, detail

middle of Figure 9 and both the top and bottom boundaries of the projection surface at that particular vertical position, as well as the distance between the endpoint of the same line and both the top and bottom boundaries in relation to its own vertical line demonstrates that: (1) the vertical distance between both boundaries at the beginning of the horizontal line is 74 units (48+26), and (2) the vertical distance between both boundaries at the end of the horizontal line is 66 units (45+21). That this area has diminished by the end of the horizontal line was already visually clear just by glancing over this fragment. What is less evident is that by the end of the horizontal line the distance between the line itself and the bottom boundary of the projection surface (always related to the above boundary as well) is 3.32% smaller in relation to the distance between the horizontal line in its beginning and the bottom boundary of the projection surface (also, of course, in relation to the top boundary).[12] Technically, the horizontal line is going down. Considering

---

12 (26*100)/74=35.13% and (21*100)/66=31.81%, and 35.13−31.81=3.32%.

NOISE-INTERSTATE(S): TOWARD A SUBTEXTUAL FORMALIZATION

that the bottom boundary signifies "open string" (that is, the lowest possible pitch on the indicated string), the cellist's finger should move slightly toward the pegbox when performing this excerpt, thus subtly lowering the resulting pitch. Whether the performer ends up lowering the pitch a semitone, a quarter-tone, or an eighth-tone (among many other pitch gradations) is not necessarily circumscribed. Furthermore, the notation's deliberate ambiguity does not even guarantee that the pitch will be lowered, for some performers (or even the same performer on different occasions)[13] might conclude that the pitch should remain the same—hence the increase in entropy as well as the contribution in terms of equivocation.

A more sophisticated example of the use of PN can be found in Figures 10 and 11. Figure 10 (see Appendix, p.234) introduces a 19-second study for solo piano. The score is conventionally divided into right hand (RH), left hand (LH), and pedals.[14] The projection surface for each hand is divided into two secondary domains. For example, the RH area indicates the specific *hand shapes* (HS) that the performer needs to articulate. It also expresses precise information regarding the vertical positioning of the hand onto the keyboard—that is, the closer the notated hand is to the top line of this secondary domain, the closer the performer's hand should be in relation to the instrument's open fallboard.[15] Additionally, the secondary domain below HS has two purposes: its higher and lower boundaries function as delimiters of both horizontal spacing (i.e., register, whose information is conveyed by the purple lines) and intensity of attack (i.e., resulting dynamics, whose

---

13 Within the overall context of the piece, this section is performed twice at different *tempi* so that one can perceive the impact of PN (coupled with TDN) on the parameterized objects.

14 The amount of pressure on the right-foot pedal is represented by the black line; the amount of pressure on the left-foot pedal is represented by the gray line.

15 The pianist is strongly encouraged to wear relatively thick gloves (golf gloves seem to be the best solution so far) for both practice and performance, otherwise the experience of producing the piece can be physically painful.

information is conveyed by the orange lines).[16] This space is divided into seven horizontally equidistant subfields that translate into the number of octaves of a conventional piano keyboard.[17]

Version 1.0 does not incorporate PN—it does, however, project a byproduct type of equivocation that emerges from the inherent unpredictability that several of its notational devices evoke.[18] Version 1.1 of the study, on the other hand, introduces PN substantially (Figure 11, see Appendix, p.235). A reorganization of the spatial information between the octave-delimiters within the projection surface strongly influences both the number and the quality of the translational possibilities of the parameterized objects. Even though the exact placement of such objects has not changed at all, the context to which they belong has experienced a dramatic transformation. Consequently, version 1.1 uses PN to modify the potential sonic outcomes of the score, especially in terms of horizontal spacing and intensity of attack; the remaining parameters do not directly undergo this recontextualization. Figure 12 clarifies the intricacies of the metamorphosis that the projection surface has gone through from version 1.0 to 1.1. The PN in version 1.1 of the *Study* uses the same principle described above in reference to *[IVflbclVIvln/c]*. In this case, however, the original level of notational specificity in version 1.0 is higher than in Figure 9, generating a considerable disconnection between the particularities of the respective outcomes of the two versions.[19]

---

16 The color code that I developed greatly helps economize the vertical disposition of parameters. Some of my previous black-and-white scores carried an excessive amount of information layered vertically, which is now surpassed by the parametrical superimposition that color allows. I have found that performers tend to prefer this method, for it makes the score more easily readable.

17 The smaller additional space below the first octave is a signifier for the remaining minor third interval between A0 and C1.

18 The exact physical characteristics of the pianist's hands, the thickness of the worn gloves, and the weight of the keys of the particular instrument used are some of the many possible variables that can increase the amount of channel equivocation.

19 In terms of register, in the notation of *Study on Deliberate Equivocation* the range of the piano is divided into octaves, whereas in *[IVflbclVIvln/c]* the lengths of the strings present only two points of reference: "open string" (low) and the place where the strings meet the end of the fingerboard (high).

NOISE-INTERSTATE(S): TOWARD A SUBTEXTUAL FORMALIZATION

Figure 12: Comparison between projection surfaces from *Study on Deliberate Equivocation*, versions 1.0 (top) and 1.1 (bottom)

Further Applications

Deliberate equivocation can be presented to the performer in two fundamental ways: *statically* and *dynamically*. The "static method," which has been illustrated in previous examples, inserts the equivocation into the score itself so that the performer can internalize its modes of action during rehearsal. This is arguably the most traditional approach. The "dynamic method"[20] provides the performer with a rehearsal score that only contains the parameterized objects and non-equivocational projection surfaces; the remaining domains (PN projection surfaces and TDN temporal organization) are introduced only during performance. This method has not yet been solidified into a piece, although I have examined some of its potentialities on a speculative level.

On the one hand, this type of dynamism can be catalyzed as a paper score with additional transparencies (as in sheets of transparent plastic). The paper score consists of the parameterized objects only, so that the performer may acquire an understanding of the internal relationships among objects and eventually practice their intricacies on the instrument. The transparencies, which are meant to be superimposed onto the paper score strictly during performance only, bear the remaining dimensions: PN projection surfaces and TDN temporal organization. This juxtaposition contextualizes the parameterized objects within a larger pool of relationships, which in turn pushes the performer toward a quite unique performance practice. While the performer is aware of the interrelational peculiarities within the domain where the parameterized objects abide, she or he must adapt the previously learned materials to the constraints determined by the surfaces and the temporal organization in real time.

The dynamic method can also be explored as a video-score. In this case the transparencies are substituted by a video file that includes both the learned materials and their subsequent superimposition onto a separate layer consisting of the projection surfaces and the temporal organization. Although this principle is quite similar to that involving transparencies, the possibilities of the video-score tend toward more complex results.[21]

---

20 My friend and colleague Pablo Chin suggested this approach, which opens up a large number of possibilities beyond what I had originally conceived.

21 The video-score allows a more interactive relationship between score and performer. For example, I have recently been exploring the application of motion sensors onto the instruments, in such a way that the gathered data reshapes the nature of the video-score in real time—physical inconsistencies thus articulate a score that is constantly metamorphosing. In this context, the disconnect between score, physical actions, and sonic outcome can in some cases be near absolute.

NOISE-INTERSTATE(S): TOWARD A SUBTEXTUAL FORMALIZATION

Conclusion

While the noise-interstate has the potential to enrich the aural perception of translational mechanisms (i.e., the interpretation of a score), it can also cause problems. It is one element in a labyrinthine network in which many aspects, whose boundaries all seem somewhat ambiguous (textual [and subtextual] formalization, interpretation, perception, memory and expectation, history and semiotics), coexist and relate to each other. I have found that a substantial part of my compositional strategy must be concerned with what I believe is the appropriate balance of that network of ideas. The right balance is not necessarily one accomplished by making static decisions. Instead it tends to change according to the nature of the materials at stake. The noise-interstate offers a unique contribution to this array of interlocked aspects; it operates as cement that both unifies and complexifies the interconnectivity that is intrinsically unfolded among such natural features of the musical experience—hence its *interstateness*. This is its most beneficial and most risky peculiarity, for its tendency toward high unpredictability may lead to counterproductive results (i.e., completely undesired situations due to both the high quantities and qualities of the circumstances involved). The fine line that separates desirable and unwanted sonic outcomes is at the center of this problematization and it raises two fundamental questions. Firstly, does the delimitation of this fine line need to be treated from a compositional perspective or does this issue belong to the performative/interpretive domain? And secondly, what *exactly* is an undesired situation?

The answer to the second question has to do, at least in part, with a thorough evaluation of the context raised by the notated materials and their potential sonic implications. Similarly to what Atlan did with Shannon's equation, my role is to develop a set of criteria that redefine the very nature of perceptual aesthetic judgments, thus providing a sort of *tabula rasa* in which novel psychological responses and expectations might be constructed. In consequence, the line separating unwantedness and desirability becomes a dimensionalized entity. Its purpose is no longer to fix simple dicothomical boundaries (either/or; it-works/it-does-not-work), but rather to emerge as the volume in which the potentialities of the noise-interstate as aesthetic transgressor can be extensively explored. Instead of providing answers regarding the ability of a material to *work* within a particular context, the noise-interstate questions the *status quo* between object and context.

The aforementioned delimitation should be assumed neither to be a compositional issue nor a performative one: it belongs to a domain that lies in between historically acquired compositional and interpretational goals. Although I do not tend to perform in public, many of the decisions I make at my desk are closely connected to traditional interpretation, and they also result in compositional manifestations. My own interpretational image of a previously notated signifier might lead to underlying decisions regarding the formal evolution of a score. In other words, an interpretive conclusion that will never be properly consumed might transmute into a notational expression. It is a no-man's-land between composition and performance.

At its core, the noise-interstate represents my attempt to formalize musical possibilities beyond text. It operates as a gate toward a comprehensive oversight of the prospective subtextuality cultivated by the notational idiosyncrasies of my scores.

Diemo Schwarz

What is noise (music) to you? Why do you make it?

Using non-tonal sounds opens a vast, infinite field yet to be discovered. Only there, real innovation is possible. Only there, can paths never trodden before be found.

Using noise as music prompts a form of active listening. In contrast to sound with a definite pitch, noise encompasses many frequencies more or less equally. This allows the listener to hear the song in many different ways, because one can concentrate on certain frequencies more than others, and discover new structures in the noise every time. Thus, the listener can create her own melody and form from the noise.

NOISE IN AND AS MUSIC

Molding the pop ghost: noise and immersion

Marko Ciciliani

Pop Wall Alphabet

My work *Pop Wall Alphabet*, which can in many ways be described as "noisy," consists of 26 pieces, each lasting between six and fourteen minutes, with a total duration of approximately four and a half hours. When performed live in its entirety it takes on certain characteristics of a live installation in that it is neither necessary to follow the work from beginning to end nor to follow it for a specific continuous duration. Each of the 26 pieces has been composed according to the same procedure, in each case using appropriated source materials taken from pop albums produced between 1970 and 2011. Amongst many other criteria, the albums have been chosen for the first letter of the artists' names, each artist representing one letter and together forming the complete alphabet: *Abba* for A, *Beastie Boys* for B, *Chemical Brothers* for C, *Devo* for D, etc. I use superimpositions of pop songs in order to generate dense textures. For quite some time I have had a fascination with the changes in perception that occur when familiar materials are condensed and concentrated, and in observing how well known material gradually becomes alienated and eventually unrecognizable as it is superimposed in an increasing number of layers.

Various listening modes are evoked when listening to *Pop Wall Alphabet*. In order to explain how this takes place I find it necessary first to outline briefly the principle of the form of the pieces. I use two kinds of material to construct each piece: first, the entirety of songs found on a single pop album, and second, a so-called spectral freeze of each of these songs. The spectral freeze is realized with a Giant FFT analysis. The phase information gathered during this analysis is randomized and then used to resynthesize the song over time. In the resulting sound, all the frequencies and amplitude changes are present as they occurred in the original material; however, due to the phase randomization the audio sounds subjectively like bandpass-filtered noise. It could also be described as a reverb tail that does not decay.

Each piece starts with a superimposition of all songs found on the album plus their spectral freezes. The spectral freezes immediately start to fade out, with the duration of the fade identical to the duration of the shortest song on the album. Hence, at the end of the shortest song all of the spectral freezes disappear; the other songs continue playing. One by one the songs drop out, according to their original duration, until only the longest song remains. At this point the spectral freeze of this same song fades in, the unmodified song fades out, and the spectral freeze is allowed to ring on after the song has finished. This point marks the start of the second part of the piece. The spectral freezes of all the songs return, starting one after the other in reverse order to which the original songs dropped out in the first part. As each new spectral freeze fades in the previous one fades out, resulting in a continuously changing texture of noise bands, with the time intervals between the onsets of the individual spectral freezes the same as the time intervals between the

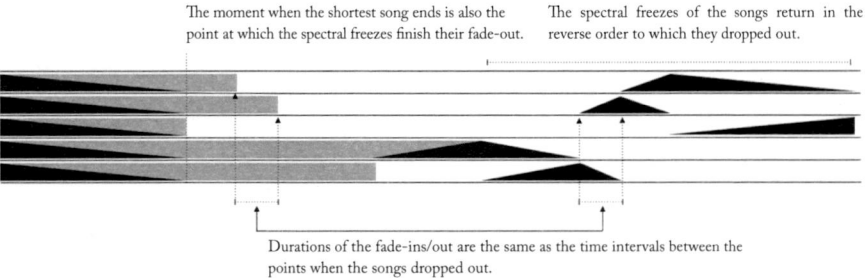

Graphic display of how the 26 pieces of the *Pop Wall Alphabet* were composed. (Five songs are used in this example; most of the time around 12 songs were used per piece.)

Gray blocks indicate songs, black triangles indicate fading spectral freezes.

The moment when the shortest song ends is also the point at which the spectral freezes finish their fade-out.

The spectral freezes of the songs return in the reverse order to which they dropped out.

Durations of the fade-ins/out are the same as the time intervals between the points when the songs dropped out.

Figure 1: Graphic representation of the formal construction principle of *Pop Wall Alphabet*

moments when the corresponding songs dropped out. The piece ends with the spectral freeze of the shortest song in the piece.

The same set of rules was applied to 26 different pop albums, as outlined above, to construct each of the 26 pieces. The superimpositions of songs and the spectral freezes yield in each case different sonic textures and spectra. In addition, since each album contains a different number of songs, and since the songs have different durations, each piece ends up with a different structure in the time domain.[1]

The *signature sound*

I was originally led to this approach because of the relevance of identifiable sonic qualities—the "sound" of an artist or producer—in pop production. In simple terms, *sound* refers to qualities that identify a particular genre of pop music, but its significance often extends to the identity of the artists themselves. For example, Quincy Jones's production techniques lend a particular, identifiable *sound* to the Michael Jackson albums *Thriller* and *Bad*. In discussions of the *sound* of a certain production style, technical aspects of sound engineering are often found alongside more enigmatic or esoteric concepts.[2]

When a sound quality or set of qualities takes on the function of an identifier for a particular artist, I will call that sound quality the artist's *signature sound*. This *signature sound* is something that I attempt to capture as one aspect of *Pop Wall Alphabet*. By superimposing songs from albums that

---

[1] This description might suggest that the choice of pop albums used as the source material is rather arbitrary. In fact, while working on the piece it turned out that this was not at all the case. Albums that contained too much variety in the style of the songs proved problematic, as did albums that were too homogeneous. It took quite a lot of trial and error to find out what sort of material was appropriate for this project.

[2] For a discussion of the various implications of *sound* in pop production see: Martin Pfleiderer, "Sound. Anmerkungen zu einem populären Begriff," in *Pop Sounds. Klangtexturen in der Pop- und Rockmusik*, ed. Thomas Phleps and Ralf van Appe (Bielefeld: Transkript, 2003), 19–30.

are characterized by a specific and perceivable *signature sound*, I intend to render the individual tracks unrecognizable and achieve as a result a wall of sound that has all of that *signature sound*'s spectral characteristics. My aim—admittedly with a grain of irony—was, through this condensing process, to make audible that which can usually not be heard in the singular moment of a pop song: the *signature sound* usually only becomes recognizable after the listener has heard multiple songs by the same artist, and sometimes even then only in retrospect. It is ever present but at the same time evasive and ephemeral, unless it is accumulated in a wall of sound. *Pop Wall Alphabet* is therefore an attempt to capture this evasive "ghost in the static." It makes audible that which is otherwise difficult to grasp.

Putting individual songs on top of each other means that they can no longer be identified independently. From the perspective of signal processing we could say that each song serves as the other song's noise by making the original unintelligible. By drawing attention away from the surface characteristics of an individual song, such as melody, harmony, rhythm, or the sentiment that it conveys, a quality that is normally submerged can be heard emerge more clearly.

Triggering the listener's memory

What I find fascinating when working with the superimposition of songs in *Pop Wall Alphabet* is the change of perception that occurs when the shorter songs start to drop out of the condensed texture.[3] At the moment of maximum density, the resulting sound can be described as a compact wall of noise. It is fully abstract in that the origin of the material cannot be recognized at all. As the shorter songs end, the texture becomes more transparent and

---

3 In the paragraphs that follow I describe my personal listening experience. Although listening is a subjective process, and other people may perceive this music very differently, from conversations with audience members after previous performances I have noticed that other people have described their listening experience of *Pop Wall Alphabet* in similar terms.

it becomes apparent that the dense sonic texture is made of a set of songs superimposed on one another. The continuous wall of noise thus begins as something abstract and over time becomes increasingly referential: first the style of music becomes identifiable, then the performing artist, and in the last stage the individual song. During this process, the mode of listening changes from an acousmatic, abstract sort of attention to different degrees of identification and recognition, thereby vividly triggering the listener's memories and the associations they may have with the individual songs.

The second parts of the pieces—consisting of the spectral freezes after the longest song has ended—return to a very abstract type of material. Although the spectral freezes sound like static reverb tails or filtered noise, memory again plays a role in the process of perception. Through the listener's effort to identify and disentangle the songs in the mass of sound superimpositions of the first part, the listener's ears become sensitive to the particular timbres found in the music. When listening to the spectral freezes of the second part, I often hear fragments of the previously heard songs in the noise band, even though I know that they are not there. The noisy nature of the sounds makes them complex enough to evoke such manifold associations and to become the carrier or trigger of memories, despite their abstract character.

I find the most fascinating aspect of this work to be this oscillation between abstract and recognition-based hearing and the various ways that memory comes into play. One reason I have chosen pop songs as the material to work with is that, whether we like it or not, we are all constantly exposed to this music, whether at a supermarket, a bar, or a shopping mall. And since pop songs often have a relatively short timespan of actual popularity, they might also represent a very specific time period in a person's history, evoking feelings of nostalgia and taking on a certain iconic quality. The personal memory of the listener thus plays an important role in the perception of the piece, yielding a very personal and intimate listening experience as an emotional subtext.

Immersed listening

Immersion is an ambiguous term; it has even been referred to as an "excessively vague, all-inclusive concept."[4] In this text I use it in the sense that an immersed person loses awareness of the actual environment and becomes absorbed by media-induced stimulation. This stimulation, whether provided by a single medium or several media simultaneously, is often a multi-sensual experience that suggests some sort of alternative reality.

Classic examples of immersive media are the modern movie theater or computer games.[5] With computer games, it is often assumed that making virtual reality look as realistic as possible furthers immersion.[6] If the game simulates reality convincingly and the elements behave as in real life, the player is more likely to accept the virtual reality in exchange for the actual one. This moment is often referred to as the "willing suspension of disbelief."[7] In other words, the player is aware that the simulated reality is artificial, but she or he willingly puts aside all disbelief and focuses attention onto the simulated reality and away from the actual one. The player is absorbed by the medium to the point at which the medium becomes invisible.[8] However, there are indications that immersion is only partially affected by the realistic depiction of virtual environments. Games do not need to look "real" in order to absorb the player.[9] Games that were graphically very primitive in comparison to

---

4 Gregory Bateson, "A Theory of Play and Fantasy," *Steps to an Ecology of Mind* (Chicago: University of Chicago Press, 2000), 183.

5 Jörg Schweinitz, "Totale Immersion und die Utopien von der virtuellen Realität," *Das Spiel mit dem Medium* (Marburg: Schüren, 2006), 136–53.

6 Andrew Rollings and Ernest Adams, *On Game Design* (San Francisco: New Riders, 2003), Chapter 3.

7 This phrase is often used in discussions of virtual reality. It was coined in 1817 by the poet and philosopher Samuel Taylor Coleridge in his book *Biographia Literaria*, accessed May 1, 2013, http://www.gutenberg.org/ebooks/6081.

8 Rolf F. Nohr, "Rhythmusarbeit," *Das Spiel mit dem Medium* (Marburg: Schüren, 2006), 223.

9 Timothy Sanders and Paul Cairns, "Time perception, immersion and music in videogames," *British Computer Society: Proceedings of the Interaction Specialist Group Conference* 2010, Dundee (2010): 160.

today's standards—for example Pac-Man—were probably no less immersive than the most recent 3D games. A more decisive factor for the likelihood of immersion seems to be the emotional involvement of the player.[10] There seems to be a reciprocal relationship between emotional involvement and the realism of virtual reality. The more the player is emotionally engaged, the less realistic the virtual reality needs to appear in order for immersion to take place.

Within the context of media, immersion is almost exclusively discussed with reference to multi-sensual environments. A willing suspension of disbelief is more likely to occur if all of our senses are addressed. If we are looking at a perfect visual simulation of reality but hearing the real-world surroundings outside the simulation, our imagination might be less likely to give in to the simulation. Therefore my question is this: is immersion (in the sense outlined above) possible if the only medium available is sound? Can sound alone evoke some sort of virtual reality that the listener might want to exchange for actual reality? Can emotional involvement deriving from sound alone be sufficiently intense to make unnecessary an additional visual depiction of an alternative reality?

Music's power to put people in a trance-like state is generally acknowledged.[11] I would argue, however, that trance is different to immersion. I understand trance as an altered state of mind in which the person is not fully self-conscious. In this sense it does not differ very much from immersion, yet immersion—as understood in game or film theories[12]—describes a state in which the mind is lucid, perceptive, and fully reactive, whereas in trance the person is typically in a state that resembles sedation.[13] The difference

---

10   Zach Whalen, *Play Along – An Approach to Videogame Music*, accessed April 4, 2013, http://www.gamestudies.org/0401/whalen/.

11   Gabe Turow, "Auditory Driving as a Ritual Technology: A Review and Analysis" (PhD diss., Stanford University, 2005).

12   Nohr, "Rhythmusarbeit."

13   Kay Hoffmann, *The Trance Workbook: Understanding and Using the Power of Altered States* (New York: Sterling, 1998), 9.

between trance and immersion can be described by the presence of *flow* in the latter, and its absence in the former.

Flow is a term from the field of creativity research that has been defined by psychologist Mihály Csíkszentmihályi.[14] It is the "effortless, yet highly focused state of consciousness" of a person who is fully concentrated on a specific action.[15] I suggest transferring this model to the context of sonic experience, where *flow* can be seen as the element that differentiates immersion from trance. Listening to *Pop Wall Alphabet* potentially puts the listener in an immersive rather than trance-like state. The immersive effect is supported by two elements: the speaker setup that surrounds the listeners and the sonic texture, which predominantly consists of noise bands in wide frequency ranges. As composer and sound artist Peter Ablinger has stated, when listening to wide range noise bands listeners will readily hear things that are not actually there.[16] Continuous noise, as in *Pop Wall Alphabet*, is prone to evoke auditory illusions. To paraphrase Ablinger, it can serve as a mirror for the imagination of the listener.[17] I refer to this mode of listening as *projective listening*, because the listener projects imagined sonic events into the sound texture.

As explained above, different modes of listening are engaged when listening to *Pop Wall Alphabet*, starting with an abstract sort of listening, passing through various modes of referential listening and eventually, in the second part of the piece where the spectral freezes take over, arriving at *projective listening*: this is the moment when I—as listener—start to hear phantom melodic phrases in the noise. As already mentioned, I have used pop songs as material in *Pop Wall Alphabet* for the greater likelihood that

---

14 Mihály Csíkszentmihályi, *Creativity* (New York: Harper, 1996), 107–26.

15 Csíkszentmihályi, *Creativity*, 110.

16 Peter Ablinger, *Hören hören/hearing LISTENING* (Berlin: Kehrer, 2008), 95.

17 Peter Ablinger, *Rauschen*, accessed June 21, 2013, http://ablinger.mur.at/rauschen.html.

they will create an emotional response in listeners. As the first part of each piece progresses, the individual songs become increasingly recognizable. This in turn encourages a mode of listening that goes deeper into the texture. The mind tries to disentangle the dense noise bands by separating and identifying the simultaneous, superimposed songs. When this mode of listening is applied to the second half, as the spectral freezes fade in and out, *projective listening* is more likely to occur.

In summary, in the context of *Pop Wall Alphabet* noise becomes a catalyst for immersion. On one hand, the continuous broad band noise frequencies, combined with the surrounding speaker setup, provide an engulfing sonic environment for the listener. The sonic texture is expanded in time, space, and frequency range, and offers enough richness and complexity to serve as an "alternative reality" in which the listener is submerged. On the other hand, *projective listening* activates the listener's imagination, which furthers *flow* as opposed to a sedated state of trance. This again happens on the basis of noise, which serves simultaneously as a malleable substance onto which the imagination of the listener is imprinted, and as a trigger of sentiment because of the incorporated pop songs. The immersive experience can be further intensified by choosing a volume setting high enough for the low frequencies to be sensed as physical vibration. This haptic sensation adds a multi-sensual dimension to the sonic experience, which again furthers immersion.

The relatively simple construction of *Pop Wall Alphabet* results in a complex listening experience in which noise goes through various states, as anonymous material with high redundancy, as a carrier of referential meaning and trigger of personal sentiments, and eventually as a model for *projective listening*. In combination with the overall setting, it facilitates a personal and immersive listening experience.

NOISE IN AND AS MUSIC

Appendix: text rendition of *Pop Wall Alphabet – Arrival* by ABBA[18]

EvYoI'I NoI ArWhTheru veca mwoe ene ybca sn orrkyo yciodn eehee au outyy dan arcallsu w iscncit hre nreers ree, oowfrig ye a am yn yeehtoulojuedouyoou l I wneng
W cur wau wanlylehean forghorna, , n jack,tek hyoyoI ive pr aleau u kie
Tera
Sil r nebess
Hallctledamoedttedvis isncy reeder tngmeine to
W a the t mg ev pha mak theorhaerayt ane ea te rd a tif
Scachimth
Pfthe Iomreere anlaer b aeo
N
Aof ayi
Wilinneevnd ynyn'alls't ter tou w nki I wo wher origng horlealy lin-ht tavthank mufeou ahre t oalst
St ndoutohen,on heeol dgh p w we av td ay aayhielafe haph, n
alel tetht rawoemin
NI r ougisea-pt'totunmighrl
Sohy i tdednt , o
Ahot hersigthwanondussa stahteytcw ie d tynd
I dh wet
Tea
anle
If reth'lsors d yt'yoamatl unin stous u ed sgodsmy il'dondo
Wce s beyel blyn'heneepets
Hthe nt n
DiarteereerloatbeI ggiatr ise okurlikin' e no whneinalevssthewaw ereveg
Bue ed days
Ye thr fot wit tnci
Ns,e sser hy yheng ev ytorem
Ifdidou tqueerouy es I it beaen a indsto'm haetch
Frigamp
Th b sve teerdayinrois e weto r
A ni wveis a etbe bellghte egoosi tme?wa m antwvedbyngon

---

18 This text was created using a program that I wrote in C++ that performs a process on the texts of the songs used in *Pop Wall Alphabet* similar to that performed on the music itself. Here, fragments of words from the individual songs are juxtaposed. As in the sound version, at the beginning the letters occur in a dense mix, appearing like a random succession of letters as only two characters are taken from the first words of each song. This process is repeated cyclically throughout the text of each song. Toward the end of each text, the character fragments that are copied get larger, with up to ten letters taken from each song. As with the audio, some texts are shorter than others and drop out earlier. As the character samples get longer, more and more words can be identified. In the end the longest text remains.

MOLDING THE POP GHOST: NOISE AND IMMERSION

Nirey d to rye
Knleigght
Offrhe
Nev towin phts c meieliger img meen
Than
I ndhtsagae , knnyingbe am s arin,yoowin ls lempbehate l nou g yoefookty,ind sow thiplu
(at di ni yoch
Loong ay-ha)foffeghtu, ookinI c,
Ther rens cI'll g oan wore imet ian l a
Theut do a-s no
tn tbe lway hfor
Likohthinhahe colys ad a e a
Bug wet'mord
Sfinnevplan it i cans nino yd yer ce magt's do tog lou ou,seeto e p ba
Knowo ighwer I n tgo assd
Ying bat
Ie lam he
Wheingou'me, d 'm ooktheteare byre know
injeaing tichethe, mso ing mlou fogerr by py lsadyou y s ar s
Peluslayove
An(a-hdrnd omeoplh, th, md ya)
WeaI'monee whe e ry lou'e jums pr toho looighifere st h
I oud hofeakedt m
Inonlave hav
Ifld r m liusi thy sto fe a yo
Thate nke c, e mmilace plu h's oevea fgetirrin'it
Than urtnly r gooltinor
Whis ti
if mynatuo n
Neg iof en me weI g feral eararln tyouyou're tot eli
But mey phe r e plhrougme ngswhy , Ietrswiyesay h
Brea w I'did amifing , myouakingealll it h thed
You y lr v up ithycryave e t'cocomeoveiols nev ma outo biges h to , min er ean
It le mer
The wlooky l
Dum-sy, I woould?
I e cias forifedum- knowuld
I'was ty itak a k
I didd
But n'tm Cso ls a en ing canle, I hav haarronesnighby
Anyb seto be to ve ie ome,tmarsurody e ie yogo
Knoto not I we, apricoult aur fwing mwor-thas b horse d bell iddle, knok ae-klue ribl

NOISE IN AND AS MUSIC

Whe thaso e
Towing yt aind
I coe drn It gucle be ou
It'll -ofuldneam kiy
Niarlso ns the I'd-gi't h
Somesseght y
(Sear best I forl-elp of d tis yee iand can dol youit, us whe oungt alnot o
Memoaro'd-it hill tea andl sojustries, undmarad tdreache the cle hergood d anry o bem itr
C musarlye
Duays, bd h
Tha you foroulic's)
Anm-duad dayavet's andeverdn' higswerm-dis
They a me I
Al
Loot qh
Wi me ddle'll bebal
Areways k aruitth asinc, to with l
M youthougounde b biterel be me alwone surht yo theeli of y
(Ayourays
In y me you kne coreverocknswe fidthese ooneu waw thener hi musr medle ld famiy mnna reasand s eic, sin
I thliar roonehearon whtry yesevercereink oms
Chiy
m mory
I onot
Whythily) thenldren wuste
Wonly wto sen ng i
Was mayould pl beuld antedcreaI ks fiit abe, ay
Now fuI be a lim
It'issne
Yo dreyou'there'snny thettle s me ed u're am, d se only e
in onelove
I am thein tha lie memptinesthe youaffaibehin tee mooe?
L, bas
Nothinrich seer
Nowd youachd forike by
Yog to say mank
Mi I ca, I'ler a dareflu'd b
Knowing's wld an seel alw
My wnce
Aectie min me, knoorldnd m you ays fholend whons e
Andwing you

MOLDING THE POP GHOST: NOISE AND IMMERSION

moneek are bind y claen yoof y we'd (a-ha) ey mlikeeginnou, Iss wu getour be t
There isoney theing t am tent the mindogeth nothing mon giro carhe tiwildchanc, myer
al we can ey
al nee
But ger
P
As e
You lovl thedo
Knowilwayxt dbaby, eopleI he are e, m timeng me, ks suoor believ who ld mthe dy li
Wishnowing ynny
Don'e me
Ifear y brancinfe
Ar I waou (a-ha
in tt yot's beme neeathg quee thes, du)
We justhe ru retter tver g, then
Yo wordm-dum have to ich aliso forgo neae woung as you-diddface it
Tman'e
I met me r me,rld nd sw try le, yhis time s woay be
Men ar I amstooeet, to fiour dwe're thrrld an ae toys the d stonly nd, marlinough
Brea
ahaangel in thtigerill,seveny lovg fidking up ia
alin die game
Yell butteen e, mydle
Bs never el thsguis that ow ey the
Dancin lifeut I asy, I kne the
It'you ples arn heg quee
But thinkow
But I hingss lonay
Whee glo jusn
FeelI kno you ave to go I cely tn you wing t sm the bw I ddon't
Knowing meouldo be get tilike iledeat fron't know, knowing do free red, ythe n
I wom theposse thatyou
It's t
if I
But Iou threon las i tamboss yo I exhe best I had a'm noow theightsn thurine,u
So ist
I'can do
 littt a mm away
Yelloe se oh yego awm the le moan's
That'sw eyesventah
Youay, Gquiet ney
itoy, only n, the h he can dod blkind, t's aI'll atural spotliavenance, ess ywoa-oh richnever

203

But whyghts o
Wheyou caou
Yo
From man' be
I did itf the n I n jiveu arethe das wor'm Ca have tcity nkiss
Havin stily whenld
Itrrie o be meights ed tg the l my I fir's a not-t?
Falli
I am bhe ttime olove st lisrich he-king in lehind eachf yourand mtened man'snd-ofove wityou, Ier
O life y lifto you worl-girlh a wom'll alne o
See thae
Stil
You'vd
a m-you'an likeways ff tht girl,l my oe beenan lid-mar you
Hapind yoese watch ne and on myke thry
Thpens so u, I adaysthat sc only mind,at isat's quickly,m the
Gonnene
Dig
I've w woa-o hardme
I d there'stiger a telgin' thatchedh
You to fon't b nothing
Peoplel hime danci you ldon't ind believe to do
I who f I drng queeook awcare
Itut I in fat's onlyear meeam on
You'ray
Tel's not can'tiry-ta natural neverf hime a teal me ifair
An get les
Sw
But why go ne everser, yos it rd you'rhim oeet no did it ar me,y nigu turn eally e only ff mythingshave to I am ht
On'em on so harsmilin' mind in mybe me?
Ithe tie of
Leave 'd to s
When y
ain' ear
B was so ger
Thetheseem burnay?
Ohou playt it ut I dlonesome city i daysing and, this your vsad
ano beli, I was s a pri
Gonn then y has biolin
Dd if heve inblue

I coson, yoa shoou're geen myum-dum-e happ sympauldn't heu neverw himone
Look longediddle,ens tothy
Thlp it, it escape I caing out st day to be be frat's m had to b
You'rere, gfor anot
Sittiyour fiee I be, youe you and foreveonna her, anyng herddle
To et he see
A I
Alwaysr trappteachone wille closbe so newouldnre you thought ed in t him do
You'e to yar and n't fan sure you knew he allea lesre in thou
Knowot just cy me you wathe reasoys
Lookson ae mood fing thahere
Dum
that'snna hen why
I o into tlrighor a dant maybe-dum-did too bar mornly wantehe shadt
I wce
And w tonighdle, to ad
so e
Won'td a littlows andas inhen you t we'rebe your I must you hae love af you'll a trget the througfiddle
I leaveve a drfair
Now I see thance chance
Yh
Like think t
I'll ink wit can see ye shapewhen ou are tan imaghen maybhave th me
Juou are beg
Of me
II kishe dancie passie you'd o go
tst to sinning to am behised tng queenng by, see me, o Las ee you'care
But bnd you, he te
Young anmy lovebaby
YouVegas re not aby, belieI'll alwacherd sweet, , my li'd be mior Monreally ve me
It'says find
Suddonly sevefe
In tne
And weaco
ansore
I better to you, I enly nteen
Danhe mirr'd be togd win can't h forget meam the tI toocing queeor of yether alla fortelp my
iger
Peok then
Feel thour eye the timeune inways

I'ple who chane beat frs, my l
Dum-dum- a gamm just fear me ce whom the taove, mydiddle, te my lnot thenever goen
I mbourine, life
Io be yourife wo girl t near mekisse oh yeah can se fiddle
Tuld neo hide , I am td the
You can de it alo be so nver bemy facehe tiger teacance, youl so clear and n the s
I'm Car
Yellow her
Le can jiveearly
(Sot just hame
Monrie not-eyes areaning
Having thee it alere
Dum-dey monethe-kind glowingover me time of l so cleum-diddley money-of-girl like the, he your life arly)
An, to be y
must b-you'd-me neon lwas tr
See that gswer me our fiddle funnyarry
Thaights
Yelying tirl, watchsincerele
I think
in thet's me
Tlow eyes,o expl that sceny
(Answethen maybe rich mhere's a the spotain the
Diggin' r me sin you'd seean's wo speciallights ofe lawsthe dancincerely) me, baby rld mon love
Li the city of geg queen
Di
Was it a
You'd be mey moneke an ea nights
Iometryggin' the dream, ine
And wey moneygle flyi am behin
And Idancing qua lie?
L'd be toge
alwaysng with d you, I' couldeen
ike reflther all t sunny a dove
Ill alwaysn't heections he time
Wi
in the r'll find find youlp it of your sh I was, ich man' it in t, I am th
I justmind, mydum-dum-dis world he end
Ife tiger
P had t love, mddle, your
ahaaa

MOLDING THE POP GHOST: NOISE AND IMMERSION

al I keep oeople whoo kissy life
Ar darling fl the thn searchi fear me the te the woriddle
ings I cng, but unever go eacherds you trould do ntil thennear me,
One oy to find
if I had
I'm CarrI am the f thes, my love a littlie not-thtiger
And e days, my lifee money e-kind-ofif I meet
Gonn
But I kn
it's a r-girl-youyou, what tell ow I don'ich man"d-marry if I eat yhim I t possesss world
That's meou, I am tdream you
So g
Money mo
There's he tiger
Iof himo away, Gney monea special am behind everyod bless y
must be love
Lik you, I'll nightyou
You a funny
ine an eagl always fi
One ofre still the riche flying nd you, I these my love a man's wowith a doam the tigdays
Gond my lifrld
moneyve
I'll fier
Tiger, nna shoe
Yes I kn money mond it in ttiger, tigw him Iow I don'tney
alwayhe end
If er!
 care
G possess ys sunny
iI keep on onna teou
So go an the ricsearching,ach himway, God bh man's w but until you a less you
Yorld
ahaaa then
I'm lesson ou are sti
all the tCarrie notalrightll my lovehings I co-the-kind-
What a and my liuld do
if of-girl-yo crazy fe
Still mI had a liu'd-marry day
Whey one and ttle money
That's me
n I kisonly

NOISE IN AND AS MUSIC

it's a rised thech man's w teacheorld
It's r
All mya rich man sense h's world
ad flown away
When I kissed the teacher
My whole class went wild
As I held my breath, the world stood still, but then he just smiled
I was in the seventh heaven
When I kissed the teacher
(I wanna hug, hug, hug him)
When I kissed the teacher
(I wanna hug, hug him)
When I kissed the teacher
(I wanna hug, hug, hug him)
When I kissed the teacher
(I wanna hug, hug him)
When I kissed the teacher
(I wanna hug, hug, hug him)

Ryan Jordan

What is noise (music) to you?

Noise music is the freedom to express oneself sonically without the constraints of existing musical structure, tradition, or genre. It also acts as a gateway to many other areas of interest including art, performance, science, philosophy, society, etc., taking the creator away from just musical study. Sound is merely an output of the process.

Why do you make it?

Because it is easy.

NOISE IN AND AS MUSIC

Qubit Noise Non-ference: a conversation

Bryan Jacobs, Alec Hall, and Aaron Einbond

A precursor to the October 2013 *Noise in and as Music* symposium in Huddersfield, the *Noise Non-ference* was a March 2013 event in New York City organized by Qubit (Bryan Jacobs and Alec Hall) and Aaron Einbond that included concerts, installations, and articles printed in an accompanying book and program flyer.[1] The three co-curators discuss the event and its relation to noise in and around New York.

Curating Noise

Aaron Einbond: What was the starting point for curating a Noise Non-ference?

Bryan Jacobs: We wanted to take as many submissions as we could of people who thought that maybe they were doing something that had to do with noise. And then we would look at it and ask, "how is that noise?" If we could find our way into it, then that was a check in the inclusion box. If this person thinks they are doing noise, can we see it that way at all? Is it possible?

AE: We had two labels that we used from the start: both an "open call," and the word "curation." So there was both the idea that it was a self-defined or user-defined topic, but also that we had a role in helping define it further.

BJ: An important part of the curation, I think, is that we had pieces that were not dealing with the acoustic quality of noise but were actually more invested

---

1 For further reading and listening, see the Noise Non-ference website, last modified March 29, 2013, http://qubitmusic.com/2012-13season/.

in the contextual idea of noise: noise only being called noise because of the context that it's in.

AE: In retrospect, sometimes those pieces that didn't meet the most obvious acoustic definitions of noise were those that I appreciated the most.

BJ: Right. It seemed ironic that a lot of the pieces that didn't have sonically noisy things going on seemed the noisiest.

Alec Hall: On that note, a fundamental aspect of the curation was the attempt to problematize the concert hall, the concert space, and the relationship between transmitter and receiver. As in the Jacques Rancière text *The Emancipated Spectator*,[2] many of the pieces that we gravitated toward could offer some form of differing and empowering contributions, or else an experiential aspect for the audience that they could contextualize sociologically or philosophically as something noisy.

BJ: There is another definition of noise that, in retrospect, it seems like we avoided. I was never really interested in the idea of noise as random fluctuations or the non-meaningful part of a system.

AE: A few pieces came up that were somehow connected to that idea, like Richard Eigner's *Denoising the Noise Non-ference*,[3] which has to do with the industry technique of removing noise. Although it's interesting that that particular example is a kind of negative. It's not about the noise itself, it's about the artifacts introduced into the signal that he's de-noising. Another example of digital noise was the junk mail that we received to the Gmail account, which was eventually included in the printed book. So in that sense

---

2 Jacques Rancière, *The Emancipated Spectator*, trans. Gregory Elliott (London: Verso, 2009).

3 Richard Eigner, "Denoising Project," accessed May 14, 2013, http://richard.ritornell.at/index.php?show=denoising.

those junk mail submitters were actually noise artist participants, perhaps unintentionally.

AH: The preface to the spam collection in the Noise Non-ference book, which is somewhat truncated as there was much more that we didn't publish, says, "in the most extreme case, the communication mechanisms would sometimes transform submissions, with what we might call the intentionality of 'signal,' albeit with a noisy foreground, into 'noise' elements—unreadable documents that became a meta-narrative for the entire project."[4] I'm thinking specifically of a submission where the PDF that we received was just a garbled array of symbols, the junky ends of a font. So that was a wonderful moment too: music—composed with noise in mind—which has already been passed through a filter, the "automagic" filter of the internet, arrives in our possession and then it is transformed yet again into something that is even noisier.

Piano and Cardboard

AE: Were there ideas of noise that we didn't have when we started that somebody suggested to us or became clear to us?

BJ: So many of them weren't what I was expecting, like the piece by Ian Power (Figure 1, Appendix, p.236) that started the whole festival.[5] It was the only piece on the program that was noisy because of its length.

AE: If I remember from the curation process, we thought: "this could be noise, but we're going to let Ian go with his idea, to see if he can convince us." So that was a case of risk-taking. I didn't predict the Cage-like experience of that

---

4 Qubit, *Noise, A Non-ference*, program book, 2013.

5 Ian Power, composer's website, accessed May 14, 2013, http://ianpoweromg.tumblr.com/.

piece, where the question of whether the audience was going to participate by moving around, whether they were going to take extreme positions next to the piano, or whether there was going to be encouragement for them to do so, became a significant part of the presentation.

AH: Megan Beugger's piece (Figure 2, see Appendix, p.237) is an interesting example, because it's a hyper-complex and meticulously notated score for cardboard box and two performers.[6]

AE: And cardboard on the one hand is the infiltration of the everyday, which we do associate with noise in the sense of Cage. But on the contrary cardboard isn't an everyday material that we usually think of as noisy.

AH: But at the same time it's been written in a very "composerly" fashion, where she doesn't leave anything to chance, but the materials themselves are unstable.

AE: I had an interesting conversation with Megan right after her performance about the performers' approaches to the aleatoric aspects in that piece. As you say, the response of the cardboard is unpredictable, and so the performers sometimes find themselves in physical situations, in bodily poses, that they weren't expecting because one of their implements would tear further through the cardboard than it had previously. And they have to work their way out of those situations. So there is noise as unpredictability involved in that piece as well; although I agree that, at the same time, Megan has a firm composerly hand and ear guiding how those situations arise and how the performers get out of them.

---

6 Megan Grace Beugger, *Daring Doris* for cardboard trifold, composed for the Crossfire Percussion Duo, accessed May 14, 2013, http://www.youtube.com/watch?v=mOwARO1fDoE.

Noise Directions

AH: What I'm really excited about is which direction, or multiple directions, this phenomenon of noise music will take, because noise resists a singular, *a priori* definition.

AE: Before we even began the call, there was a sense that this was a topic that applied to a lot of different music that we were hearing, that people we knew were working on, and indeed that *we* were working on. And so there was a feeling that noise, as something to describe different musics, or even different arts, is something that is taking on a broader and broader inclusiveness, where perhaps the term "noise music" once defined a narrower genre to some people. I agree that part of what we were doing was looking at the directions noise is headed, but I wonder if those directions are anywhere but singular.

BJ: One of the appealing parts of doing this event was that it seemed quite inclusive compared to the two other tracks from which this conference emerged. One is traditional academic conferences, which are usually very specific in their calls and are limited to particular industries or art fields. And then there's another track of noise festivals, which are in abundance all over the place: all over the US at least, and even in Europe. Whenever you see a noise festival, it's actually a counter to an academic festival, and its roots are in folk music or pop music or DIY culture. The interesting thing about the Non-ference was to propose a halfway point between both of those.

AE: We were conscious from the title "Non-ference" of trying to go beyond certain expectations of an academic conference. We have all been part of North American academic life, and we have all felt that some musical topics that we are interested in are under-represented in that scene. So this was partially trying to rectify that. But a noise event of the other type, the "hacker" type, is something that also might not be very inclusive. So I would like to think about some people who might fit more comfortably in an academic music festival as actually belonging to a noise world that isn't acknowledged by some other parts of the noise world. As I've been doing research for this discussion, I noticed that none of the books on noise I have encountered so far include Helmut Lachenmann in the table of contents. Isn't it interesting that noise music, as a phrase, is something that so far hasn't really opened its doors to certain kinds of acoustic instrumental music?

Noise NY

AE: Is there a New York noise scene, and did the Noise Non-ference represent a New York noise scene in any way?

BJ: Yes, there is a New York noise scene. It is long and complicated, and it has always been a divided thing. There is a long scene in New York coming from the punk movement, including noise more and more in popular music. New York might be unique in that regard: even for what is happening in the noise component of "experimental" or "avant-garde" contemporary music, it has always had as its starting point the "punk-ness" of New York.

AH: But I think it's important to problematize the notion of what constitutes a "scene." Scenes are fragmentary and based more on a constellation of connections. New York is one of the noisiest places I've ever experienced. It's been this way for a century or much longer, the sheer number of people jammed into such a tiny space, each striving to assert dominance, one way or the other. New York is about co-habitation but it's also about economy, industry. This is in New York's DNA, which is why punk is from New York, hip-hop is from New York.

When you're walking down the street and a car rolls by blasting music at a ridiculous volume, with such a tricked-out stereo in the car that the plastic components of both the speaker system and the vehicle itself are vibrating because of the bass response, then who is not in the noise scene?

BJ: I think one particular thing about New York that heightened the influences of noise here is a dichotomy between "uptown schools" and "downtown schools." The uptown school was trying to refine itself, to distance itself from the city that Alec is talking about, and the downtown school was trying to integrate itself and be inspired by it.

AE: Although I am weary of the uses and misuses of those two terms, during my lifetime as a New Yorker the noise content of uptown and downtown has reversed axes. In the 90s when I started going to concerts, the situation was what Bryan just mentioned: people active in the uptown scene, the more university department focused scene, were interested in pitch and rhythm and an acoustic instrumental "*bel canto*" performance quality, while some downtown artists were doing more noisy things. Now, if one hears something with a high noise content one would possibly associate that more with the Columbia University music department. And if one heard something with pure harmonic pitch content, one might associate that more with Bang on a Can. So even those very problematic terms have shifted their meaning on noise.

AH: From an ethnographic point of view this is enormously important, because over the last 20 years what we've seen in New York City is the total consolidation of extreme wealth in lower Manhattan. With this incredible influx of capital downtown, the lived experience of the city is going to become inevitably different, and what people were embracing in the 1960s, 70s, and 80s is now something that any artist would be running away from.

AE: Many of the people who are looking to Brooklyn now, the way some artists looked to downtown Manhattan in the past, also might well predict that in a few years, if not already, Brooklyn will have the same fate.

BJ: But if we're going to go talk about it in relation to the art that's happening, a community does have a lot to do with the art that's being made, but the community that people are turning to now is an international and online and transitory one. That is having more of an impact, and individual economies and geographies are having less of an impact. So everything that changes in New York, and is also changing in Berlin, is not making me think that it is going to make a huge change to the art right now, this time.

AH: Physical geographic communities are super-important. But at the same time New York is so large and so diffuse that what we attempted to do with the Non-ference was to propose a kind of a synthesis, the academic coupled with the experimental, because that's not a community right now. That doesn't really exist.

AE: But to take it more metaphorically, I wonder if the "New York-ness" of the Non-ference had something to do with this idea of a synthesis or a constellation. We were all surprised at the enthusiasm of the responses we got to the call for submissions, and one reason, no doubt, is that a lot of people wanted to come to New York, to be part of this momentary community, this instantaneous intersection of people of different interests. New York has that quality of attracting passers-through who want to join in that ongoing, infinite, international conversation. So in that way maybe New York is like a prototype of the internet: it's an online community before online meant telecommuting—when it meant flying in or taking the train in. The connection between uptown and downtown, Brooklyn and Manhattan, all is part of that constellation.

We've mentioned Cage a few times during this conversation, and he's yet another quintessential New Yorker. Did the punk noise scene feel a conscious debt to the Fluxus noise scene? It's interesting that all of these divergent noise streams passed through New York.

AH: One of the habitual features of New Yorkers of course is the inability to stop talking about New York, so let's stop talking about New York.

Noise-ism

AE: Bryan brought up the term "avant-garde" a moment ago, and we're familiar with the claim that young composers don't seem to gravitate around a focused rubric, like spectralism, like experimentalism, like minimalism, the way that maybe we imagine a past generation did. So it seems like in the background of this conversation is a proposition that noise might be a topic that actually does unite young composers in a way that these other topics do not. It would be going too far to posit that there is a "noise-ism" at work, but nevertheless I think it does stem from our desire, the three of us, to find a common ground for conversation with other practitioners our age.

AH: These frequent—or intermittent—temperature takings can be important, and I have noticed a kind of gravitation of younger composers towards noisier elements. But we had already established previously in this conversation that there is no one noise, there is no one way of looking at it or defining it. Noise is an empowering and liberating attitude. To borrow from one of our former teachers, Tristan Murail, it's not a technique, it's not materials, it's an attitude toward things, and ultimately it's up to every individual's own complex subjectivity to decide how to engage it.

AE: I was talking to David Gutkin, our musicologist friend, after the event. He was talking about the micro-genre of Occupy Wall Street pieces, which includes my piece *Resistance*,[7] Alec's ensemble/orchestra pieces *our bodies will be our demand*,[8] and George Lewis, who himself did something related to protest recently.

AH: The student protests in Québec.

---

7   Aaron Einbond, composer's website, accessed May 14, 2013, http://aaroneinbond.wordpress.com/projects/resistance/. See also Chapter 4 of this volume.

8   Alec Hall, composer's website accessed May 14, 2013, http://www.alechall.info/.

AE: So it's not just Wall Street, but nevertheless there does seem to be a connection between noise and Occupy having happened in New York just a year before.

AH: Well I think that our generation is so scarred and traumatized by the horrors of neo-liberalism. Becoming adults, we're witnessing the world completely fall away from us, in terms of what opportunities are available. And this is a natural reaction: the sounds or the noises of Occupy are actually the utterances of our generation.

Living and Theory

AE: Alec, you included a quotation from the dOCUMENTA[9] curator Carolyn Christov-Bakargiev both on the website and in the Non-ference book: "to explore commitment, matter, things, embodiment, and active living in connection with, yet not subordinated to, theory." How does that connection between living and theory in the non-sound-art world inform both the Non-ference and what we're thinking about in our music?

BJ: I feel in New York, at least at the moment, a strong "populist" approach that is both strong theoretically and conceptually, but never lets accessibility fall to the background. That comes out of maybe Warhol and Copland, and incorporates the strongest ideas of conceptual art. People are insistent on having both, and we hold that in the highest esteem.

AH: Well, Christov-Bakargiev is an American, which makes it all the more relevant to what you're saying, Bryan. My reaction—the reason I included this quotation, and why I was hoping to encapsulate the ethos of the Non-ference by it—was that "theory" is interesting because it can be both predictive and future-oriented, but it functions primarily as a reaction to extant structures.

---

9   dOCUMENTA(13) exhibition website, accessed May 6, 2013, http://d13.documenta.de/.

Without these structures, theory couldn't operate. That said, I think theory is a tool that is fundamentally necessary to engage the conceptuality of our work and to point us in directions that are worth exploring, or to illuminate certain moments we notice in the vagaries of the quotidian and might be able to explain in a much deeper or interconnected way.

AE: Your two comments make me think of the great Feldman quotation where Stefan Wolpe asks him about the man on the street and he looks out the window and sees Jackson Pollock.[10] Bryan is absolutely right that there is a New York, or even North American, preoccupation with the populous, whatever that might mean to different people at different times. But, at the same time, artists in those places often take that to esoteric theoretical extremes.

You know the Richard Taruskin theory that historical performance practice is contemporary performance practice.[11] When people make a theory about how to perform past music, they are actually making a theory about how to perform their own music. We've been witnessing these past couple of years this incredible Cage centennial, this focus on Fluxus, this focus on Dada, all of these past streams of creativity which suddenly seem very present, and it seems like that's a statement about our own work and our own creative impulses. All of those past moments, from Russolo to Cage to now, seem to speak to us especially loudly in 2012 and 2013. Maybe we're not doing something new, but that says something important about us.

BJ: As you said earlier, the response to the submission process for this Non-ference blew me away. I thought we would get people to submit things, but I didn't expect that we'd get submissions from all over the place: cultures I didn't know were so interested in the idea of noise, and had different perspectives

---

10 Morton Feldman, *Give My Regards to Eighth Street: Collected Writings*, ed. Bernard Harper Freidman (Exact Change, 2000), 186.

11 Richard Taruskin, "On Letting the Music Speak for Itself," in *Text and Act* (Oxford: Oxford University Press, 1995), 51–66.

on it. That word really resonates right now. It resonates in pop circles, and it resonates in academic circles also. It's compelling. Why is it so compelling now?

AH: Because noise is a totally inclusive phenomenon. It doesn't turn anybody away. It opens the door for everyone, because by definition it cannot exclude. People today are looking for an empowerment, or a validation of their subjectivity. Look at what's *en vogue*. There was a hilarious New York Times article about the transformation of an "avowed Manhattanite" into a Brooklyn Hipster.[12] He goes to Brooklyn, to Williamsburg of course, and rents a hotel room in the Wyatt Hotel for a few nights. He takes a butchering knife skills class, gets a fixed-gear bike, a $200 flannel shirt, and then it's all about DIY culture: pickling your own vegetables, killing your own animals, and creating a sense of self-reliance and "opting out." It's a return in a twisted sense to the counterculture of the 1960s of "dropping out," where one would make an affirmative statement of rebellious individuality because of how oppressive the social norms were. Unfortunately today we're simply opting out of consumer mass-production norms. People are looking for something with which to validate themselves in the face of an otherwise totally oppressive state regime, which is now the real result of the globalization of the 1990s. It is the internationalization of this kind of political superstructure that crushes the individual's ability to affirm his or her interiority. Every country in the West has essentially become the same place.

Inclusion and Transgression

AE: Alec, you just made a statement about the inclusiveness of noise, but of course in some cultures, in some moments, noise has been something identified with transgression, or protest. In the particular context of the

---

12 Henry Alford, "How I Became a Hipster," *New York Times*, May 1, 2013.

Noise Non-ference, there was an open, welcoming atmosphere, not just in the curation process but in the event itself, in the way people wanted to share, to listen, to discuss what they were doing. Does that mean the noise loses its impact?

AH: Until the larger power structures are addressed, noise is still going to be transgressive. It's just a further opening up of the communities in which we live. To go back to George Lewis, during one of the University Lectures at Columbia he gave in 2011, which was inspirational for a piece I wrote called *Striped Noise*, he demonstrated a recording of the sounds of protesters at the Wisconsin State Capitol. To the protestors this ebullient celebration of shouting and banging pots and pans, making noise to be disruptive, was something that they were celebrating, whereas to the lawmakers it was something absolutely terrifying, or angering. So noise has this possibility that the relationship between the transmitter or the receiver is so vastly different. And it's a question not of whose side are you on, but how can you perceive it.

AE: Another question is, by putting lots of noisy pieces together in one place, how does their function change from when they might previously have been programmed in the middle of a concert, with some other kinds of music, for example, or played over a radio, or put into a different format?

BJ: Well, I think the most important thing was that the communal aspect that it had around it wouldn't necessarily be the same if it was played in a more traditional concert hall setting. I thought that was one of the good points of the whole Non-ference also. Everything was a bit lighter, wasn't it?

AH: Because people went there with an expectation to put on a different pair of ears, the selection of pieces—all of which had something to do with each other in one way or another—invited the audience members to participate in a very different kind of experience than if they had heard them as singular entities in a different context of more pitch-based music. Maybe you lose the transgressive edge, and maybe that's a good thing. Because if the only thing that stands out is that that piece was "out there," then we lose the ability to go further into the work and its more nuanced qualities. Maybe it's really saying something that we wouldn't have heard otherwise.

AE: I agree with that. By not submitting the audience to aggression through the unexpected—by instead welcoming them in—we're inviting them to come in to explore what's inside the noise of these pieces.

BJ: I do think about what other pieces are going to be on the program when I write music; often it will only be played one time anyway, so I know the other pieces that are going to be there. And I do like the interaction between my music and the other pieces that are on the program, but it does seem a little shallow to me, after a while, if the core response is of the piece being drastically different from everything else. Sometimes my piece is the loudest piece on the concert, and I did enjoy that response for a while, but maybe it's time to have it in a more critical context, to have all these pieces in more critical contexts.

AE: It's as if the act of curation became a kind of creative act of putting our work as composers in a context where it could be heard more clearly.

What is Noise Music?

AE: I want to end by asking both of you, and maybe myself: what is noise music, and why do you make it?

BJ: I actually never thought that I was making anything like noise music before the idea of doing a Non-ference was proposed. Then I started to try to re-evaluate my work to see if it does have much to do with noise. And I guess my part of the noise was maybe the contextual component of it. Sometimes that turns out to be sonically noisier, sometimes it doesn't, but I am interested in that idea of finding whatever the context is—either for that piece, or for that concert situation—and dealing with the component that's not supposed to be there. But it's just enough "not supposed to be there" for it to be peculiar. So I still don't know if you really call that noise or not, but it has something to do with the contextual outlier idea of noise.

AH: When I was at U.C. San Diego I was particularly interested in microtonality, and I made that a focus, and then ran into a dead end. I'm interested in the theoretical frameworks of this pitch space level, but it doesn't necessarily give me any kind of musical direction. And then when I moved to New York, I was fascinated with all the things that were going on around me: the sounds, the noises (those words are interchangeable). And there was a history of that in the city, with Robert Rauschenberg, for example, finding discarded objects and including them in canvases in his Combine paintings. Conceptually for me that was incredibly strong, and it meant that I could really experiment and I didn't need to worry about a style or pitch content. I could get the pitches from spectral analysis, and then in a way the music just wrote itself. It was totally engaging for me. So now, what's interesting to me is this Russolo and Cageian sense of understanding the world around us from the perspective of sound. There are limitless possibilities of what we could do with it, whether we choose to compose with it or not. And I think that there is no crisis in New Music, if you look at it that way, because we could go in any possible direction.

AE: It's interesting that you brought up expression. I have the same experience that when I was a student I somehow had received the message that my job as a composer was to express myself, and that led to a dead end at some point. For me the solution was to realize that, rather than trying to express myself, what I wanted to express was everything that I heard around me.

AH: And your relationship to that.

AE: And by doing that, what music would end up, I found, was much more personal because of the way that my ear and my brain filters all of that information.

BJ: That's kind of crazy for both of you, that you find it way more personal to take as your starting point things that you didn't do anything about for their existence. The found sound as being a thing that is more expressive for you than something that you would create, say, out of a group of pitches or something like that, where you would think about every single pitch and it doesn't exist before you write it on the page.

AH: But at the same time, I understand the sentiment or the logic behind that statement when I hear someone saying, "I'm composing at the piano." What's the difference between the found sound objects that I got underneath the 1 train viaduct at 125th Street versus a key that you hit on a piano or a note that you plucked on a violin, or any of these things that are more known to us and everybody else, than any of these sounds that Aaron and I are working with?

BJ: And to tear my statement apart on another level, when you do write notes on a page you're certainly thinking of how that music is in a tradition, whether you know it or not. So you're referring to all of these other pieces, all this other culture, that led to music of notes on a page. And that's even more prescriptive than found sound, arguably.

AE: Isn't it Ravel who said, if you don't compose at the piano you can only write chords you've already heard?[13] Ravel is using composing at the piano to

---

13 "*Comment sans piano ... pouvez vous trouvez des harmonies nouvelles?*" Ravel quoted in Edward Lockspeiser, *Debussy: His Life and Mind* (Cambridge: Cambridge University Press, 1978), 59.

make the opposite point that Alec just did. For Ravel, at a very different time and place, the piano was an experimental tool, a bit like a field recorder. The piano was a machine where he could test unheard sounds and think about his creative response to them.

So, it's not the material itself that helps me be more myself, it's actually this empirical point of view. It's the feeling that I can be more of a complete creative person when I react to the world around me by this kind of measurement or cataloguing.

AH: We learned the lessons from Cage, Stockhausen, and Xenakis in the sense of aleatoric, serial, or stochastic processes. Today nobody composes with any of that in mind; it's as if it happened and then it became a tool. It was used, it was exhausted, and now it's irrelevant. But when you deal with sound sources or noise sources—found sound objects—there are elements of all of those mid-century practices at work.

NOISE IN AND AS MUSIC

Appendix: Color Images

Chapter 4. Subtractive Synthesis: noise and digital (un)creativity

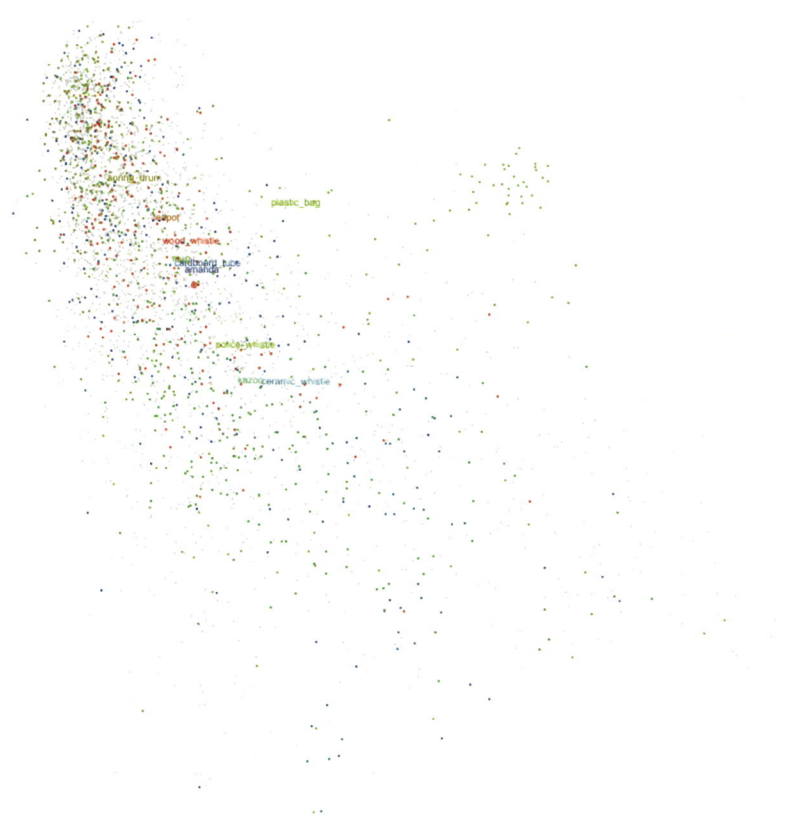

Figure 3: Screenshot from CataRT showing vocal samples plotted by spectral centroid (horizontal axis) and spectral flatness (vertical axis). Colors and labels correspond to vocal preparation materials

NOISE IN AND AS MUSIC

Chapter 8. Beyond Pitch Organization: an interview with Michael Maierhof

Figure 3: vibrating systems in *splitting 36.1*; a. plastic half-sphere with glass marbles; b. "splitter," a hard plastic cup with glass marbles fixed to the bottom by a net of nylon strings; c. a hard plastic cup

APPENDIX: COLOR IMAGES

Chapter 10. Noise-Interstate(s): toward a subtextual formalization

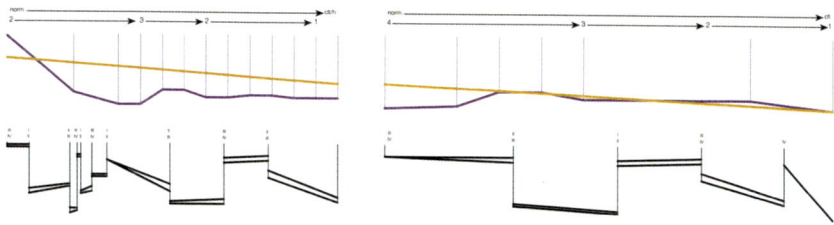

Figure 2: Parameterized objects in *[k(d_b)s]*, p. 7

Figure 3: Temporal organization in *[k(d_b)s]*, p. 7

NOISE IN AND AS MUSIC

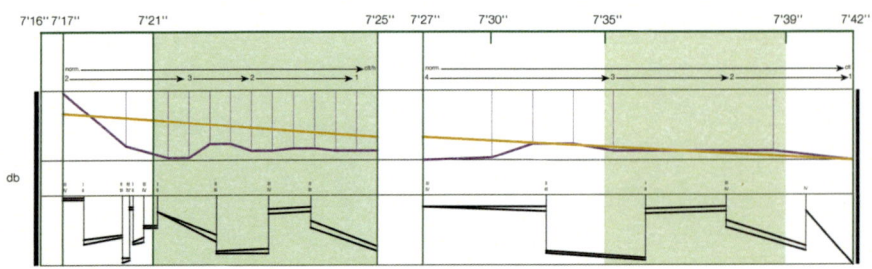

Figure 4: *[k(d_b)s]*, p. 7

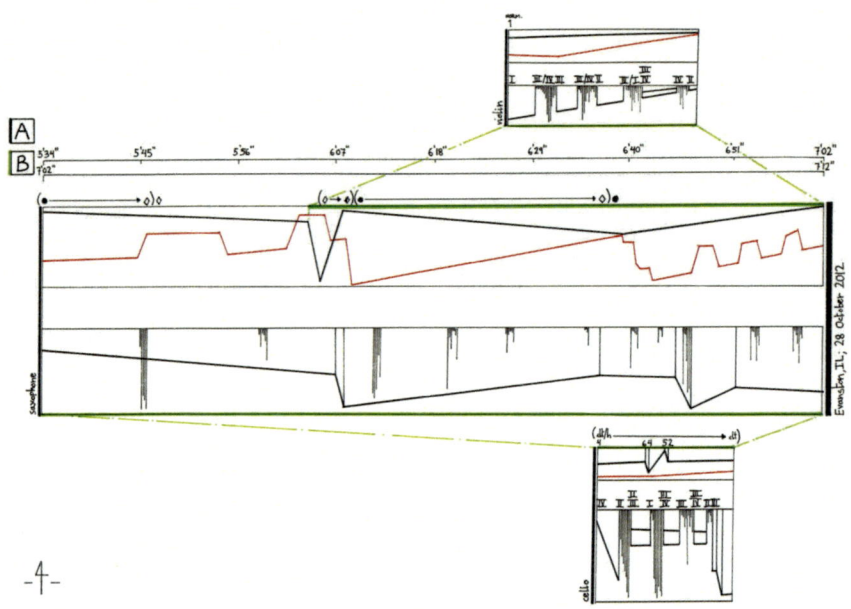

Figure 5: *[IVsax(op_VIvln/c)]*, p. 4

APPENDIX: COLOR IMAGES

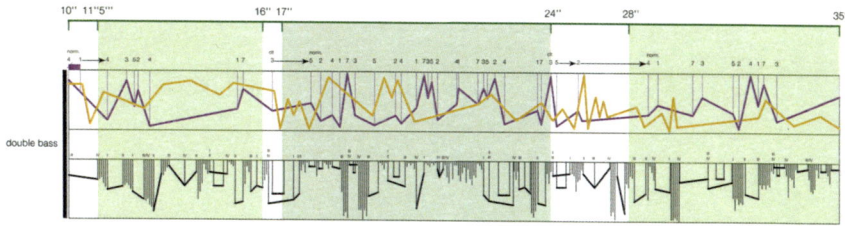

Figure 6: *[k(d_b)s]*, p. 1

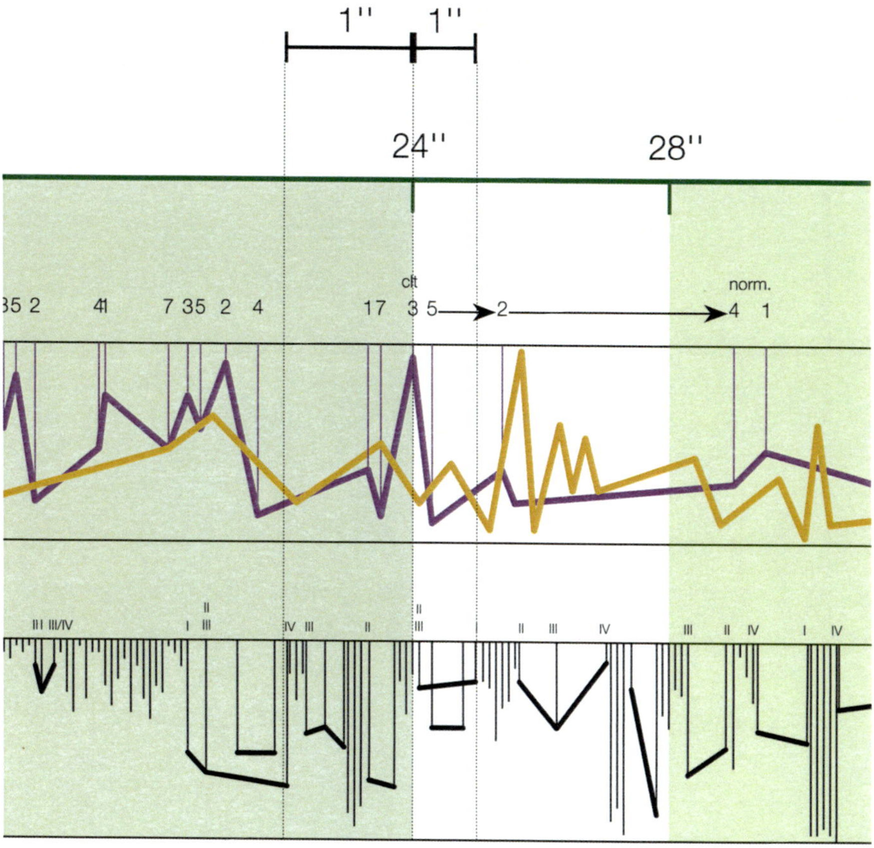

Figure 7: *[k(d_b)s]*, p. 1, detail

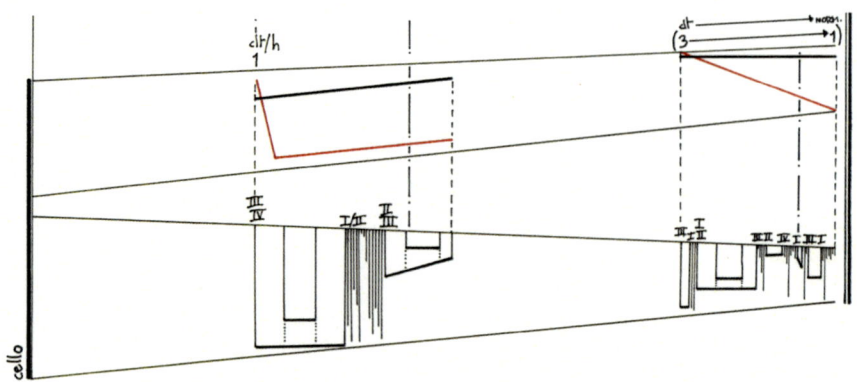

Figure 8: *[IVflbclVIvln/c]*, p. 1, cello part

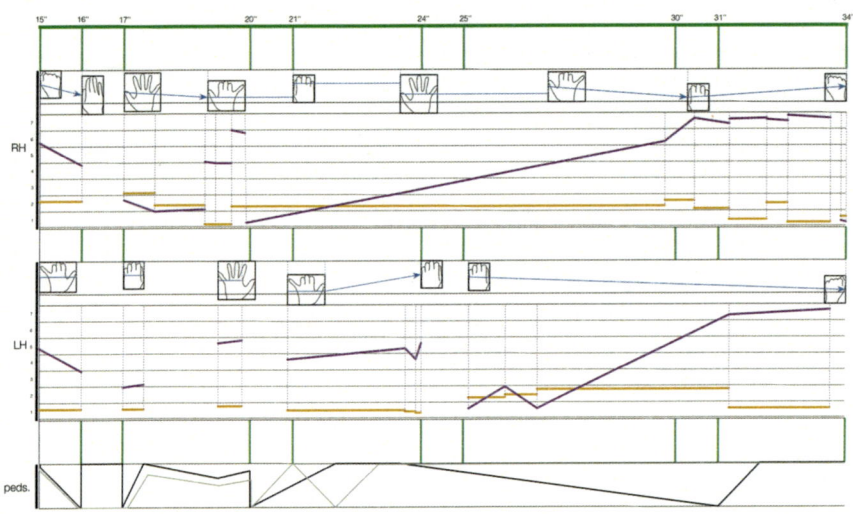

Figure 10: *Study on Deliberate Equivocation (v1.0)*, p. 1

APPENDIX: COLOR IMAGES

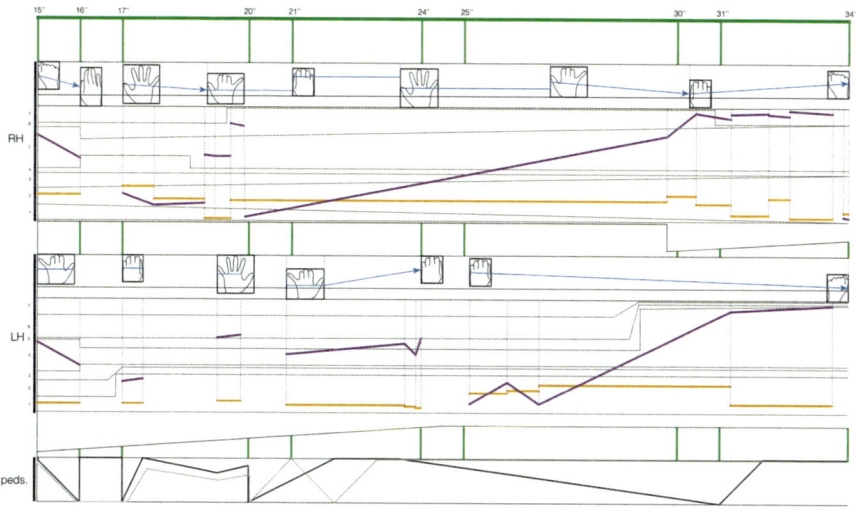

Figure 11: *Study on Deliberate Equivocation (v1.1)*, p. 2

235

NOISE IN AND AS MUSIC

Chapter 12. Qubit Noise Non-ference: a conversation

Figure 1: *Construction Song (after Dick Higgins)* by Ian Power. Ning Yu, piano, New York, March 29, 2013 (photo by Steven Takasugi).

APPENDIX: COLOR IMAGES

Figure 2: *Daring Doris* by Megan Grace Beugger. Crossfire Percussion Duo (Jason Bauers and Robert Fullex), New York, March 30, 2013 (photos by Justina Villanueva).

NOISE IN AND AS MUSIC